The Structure and Reproduction of Corn

50TH ANNIVERSARY EDITION

T.A. Kiesselbach

The Structure and Reproduction of Corn

50TH ANNIVERSARY EDITION

T.A. Kiesselbach

Cold Spring Harbor Laboratory Press
COLD SPRING HARBOR, NEW YORK

This Cold Spring Harbor Laboratory Press edition is a reprint of Research Bulletin 161, University of Nebraska College of Agriculture, Agricultural Experimental Station, November, 1949. A note from the publisher and an index have been added.

The Structure and Reproduction of Corn
50TH ANNIVERSARY EDITION

Printed in the United States of America

Project Coordinator: Elizabeth H. Powers
Production Editor: Dorothy Brown
Desktop Editor: Susan Schaefer
Cover Designer: Ed Atkeson

Cover: This most beautiful and accurate drawing of the maize plant is reproduced here in simplified stamp form with permission of its author-artist, Walton C. Galinat.

Library of Congress Cataloging-in-Publication Data

Kiesselbach, T. A. (Theodore Alexander), 1884–1964.
The structure and reproduction of corn / T.A. Kiesselbach.
p. cm.
Originally published: Lincoln : University of Nebraska, 1949.
ISBN: 0-87969-556-0 (cloth : alk. paper)
1. Corn--Anatomy. 2. Corn--Reproduction. I. Title.
SB191.M2K5 1998
633. 1'5--dc21

All Cold Spring Harbor Laboratory Press publications may be ordered directly from Cold Spring Harbor Laboratory Press, 10 Skyline Drive, Plainview, New York 11803-2500. Phone: 1-800-843-4388 in Continental U.S. and Canada. All other locations: (516) 349-1930. FAX: (516) 349-1946. E-mail: cshpress@cshl.org. For a complete catalog of all Cold Spring Harbor Laboratory Press publications, visit our World Wide Web site: http://www.cshl.org.

Contents

A Note from the Publisher

Earlier this year, a conversation with Ueli Grossniklaus, a plant geneticist at Cold Spring Harbor Laboratory, brought to the attention of the Laboratory's Press a publication entitled *The Structure and Reproduction of Corn*, by Theodore Kiesselbach. This book, it seemed, was much sought after in plant biology labs throughout the world but could now be found only in the form of smudged and dog-eared photocopies. Intrigued, we began inquiries about the origins of this work and quickly discovered that Kiesselbach had been a professor in the University of Nebraska and his book had been published by that University's Press in 1981. It was long out of print and we began considering the acquisition of rights to a reprint. Then, quite out of the blue, came a message from an old friend, the distinguished editor and publisher Sarah Greene, who understood, she said, that we were interested in a book written by her grandfather. Through Sarah, we discovered that the 1981 book had originated in an agricultural research bulletin published by Kiesselbach in 1949, and we were able to obtain an original copy of that bulletin from which to create a new edition, as well as a photograph and other material, including fragments of the author's unpublished autobiography, that illuminated the person behind the science. It was a particular pleasure to discover in this material a direct connection between the man who was one of Nebraska's agricultural heroes and Cold Spring Harbor Laboratory.

Corn science at Cold Spring Harbor will forever be associated with Barbara McClintock, the iconoclastic investigator whose painstaking observations of maize development first disclosed the activities of transposable genetic elements. But McClintock was not the first scientist to make dramatic discoveries in the corn fields at the head of the Harbor. That honor goes to George H. Shull, one of the first members of the Station for Experimental Evolution founded by the Carnegie Institution of Washington in 1904.

As an enthusiastic Mendelian, Shull was interested in the analysis of quantitative characters. Because corn is wind-pollinated, not all of the 300–500 kernels in each ear are fertilized with pollen from the same plant, so they may differ widely in easily measured traits such as size, shape, and number. Shull chose to measure variability in the number of kernel rows in an ear, and, substituting hand pollination for natural fertilization, he embarked in 1905 on the creation of inbred lines of corn. As breeding progressed, the vigor of the lines declined: The number and size of ears dropped and the plants fell prey to disease. But when these lines were crossed, Shull observed that the progeny were superior not only to their parents, but also to the original, open-pollinated corn plants from which the lines were derived. Similar findings were made independently and simultaneously by Edward East at the Connecticut Agricultural Field Station.

In the late 1800s, corn had become a crop of supreme agricultural importance in the United States. Between 1866 and 1900, the acreage of corn tripled and production increased fourfold, so that by the turn of the century, twice as much corn as wheat was being produced. As a member of an institute for basic research, George Shull began his experiments not to seek practical agricultural applications but to test a criticism of de Vries's work in *Oenothera* that mutations were an artifact of selfing a species which was naturally cross-fertilizing. Nor were he and Edward East the first to demonstrate that inbred strains lacked fitness and that crossing such strains promoted an increase in vigor. Shull's truly original contribution was to suggest that since corn plants could not self-fertilize, and every plant in a field of corn was naturally a hybrid resulting from random fertilization, the process could be controlled for the grower's maximum benefit. He proposed a "pure line method" of corn breeding in which otherwise useless inbred lines were maintained for the benefits that accrued from crossing them. The corn breeder, he maintained, should be concerned not with finding the best lines but with finding and maintaining the best hybrid combinations.

The technique Shull suggested produced seed corn that was too expensive for large-scale use and it required the invention of the double-cross method of breeding by East's student, Donald Jones, before hybrid seed became commercially viable. Nevertheless, Shull, who neither trained in an agricultural institution nor ever worked in one, became closely involved with practical issues and spoke to horticultural and agricultural researchers at conferences and on campuses all over the United States. Then, as now, advances in the manipulation of crop species had their detractors, and at the 1909 Corn Exposition in Omaha, Nebraska, Shull made an impassioned defense of hybrid corn, urging his mid-Western audience to overcome their antipathy and examine the value of this new approach to breeding for their farm practice.

In the audience that day was E.G. Montgomery, Professor of Plant Crops at the University of Nebraska, and his young assistant, a recent graduate in his first job, whom Montgomery was paying 15 cents an hour to do data entry. Theodore Alexander Kiesselbach was a Nebraska native whose mother had moved her few surviving children to Michigan after the deaths of her husband, eldest son, and father. They returned to Broken Bow, Nebraska, in 1895 when Theo was eleven, and moved into his uncle's farmstead. Although their new timber-framed home was a cut above the neighbors' sod houses, it lacked enough rooms to provide Theo and his siblings with somewhere to sleep. But it did have three empty grain bins, which were tidied up and served as bedrooms for the boys. And at harvest time? In an unpublished autobiography completed shortly before his death in 1964, Theo remembered learning to sleep on blankets spread out on top of wheat piled almost to the roof and to avoid sudden descents down the ladder.

The Kiesselbachs moved next to Lincoln, Nebraska, so that the boys could attend high school. There Theo attended a church in which Sunday School was taught by William Jennings Bryan, already a nationally known orator, who in 1896, at the age of 36, and again in 1900, had run unsuccessfully for the Presidency. Theo later recalled that the topic for discussion one Sunday was "Success." Sensitive to the feelings of his teacher, Theo charitably took the view that success should be judged by the effort expended rather than the goal achieved. But Bryan would have none of it: a meritorious objective must be achieved, he maintained, before success could be bestowed. In later years, Theo acknowledged, he came around to Bryan's way of thinking.

Bryan later achieved distinction of another kind, as the successful prosecutor of John Scopes and the defender of creationism in the famous "Monkey Trial" of 1925. Theo's own position on Darwin's ideas was tested early, in high school, when he presented a senior thesis entitled "The Theory of Evolution." Far across the country, at the field stations of Cold Spring Harbor and Woods Hole, Darwinism may have been exciting discussion and experimentation, but these ideas were still far from universally accepted. It took courage for the 18-year old Theo to write in 1903 that "probably to a large degree, the modern ethical standards of man have arisen as a protective element of survival. Some of the desirable virtues now dominating within the more advanced civilizations are: love, justice, integrity, honesty, industry, consideration, respect, moderation, search for truth, decision, monogamy, self-control, faithfulness, helpfulness, sincerity, and generosity. Based on my observations, I would recommend to every person a general blending of all these virtues as a code of ethics, or religion if you wish to call it that." Dr. Wolfe, the principal, told Theo his thesis was splendid but that the organizing committee had decided against asking him to present it at the forthcoming commencement. Evidently the young Darwinist was judged too controversial for this very public forum.

From high school, Theo went on to the University of Nebraska and graduated with a diploma, acquiring a good broad general education but no specialized training. Intending to marry soon after graduation, he wanted a real job, with prospects. His dream was to take up the tenancy of the farm that had been his boyhood home, but his future father-in-law maintained that his bride-to-be was too frail for farm life and advised an urban career. After a desperate search, he found the position with Professor Montgomery, who encouraged the young man's ambition and helped him enter graduate studies in botany and plant breeding.

Inspired by Shull's address, Montgomery and Kiesselbach immediately began inbreeding the best varieties of open-pollinated corn available, and in 1913, they produced the first corn hybrids grown anywhere west of Connecticut. Building on this advance was a slow and frustrating task. Hybrids frequently had structural weaknesses that affected their viability, and drought wiped out the breeding stocks. Only in the 1930s was hybrid corn truly established in the state, but by 1949, it accounted for 96% of all corn grown in Nebraska, boosting the income from the crop by $42 million a year more than the earlier stocks had yielded.

Kiesselbach went on to an honored career in the University of Nebraska, becoming Professor of Agronomy and throwing himself into efforts to improve the lot of Nebraska farmers. He was instrumental in the establishment of new crops such as Sudan grass, sorghum, soybeans, and clover and led vigorous resistance to overhyped claims for Jerusalem artichokes and other fads. The author of more than 70 research papers, 60 research bulletins, and innumerable articles for the agricultural press, his interests in crop science ranged widely but remained centered upon the corn plant.

At the 50th anniversary of the first publication of *The Structure and Function of Corn*, plant science is still an important focus of research at Cold Spring Harbor Laboratory, and the corn plant remains a valuable subject of study, now with the powerful tools of molecular genetics. The Laboratory's Press takes great pride and pleasure in restoring to the plant science community a reference work of lasting value, in a form that displays its text and splendid illustrations to best advantage, and we honor its author, Theodore Kiesselbach, as one of the pioneers in corn genetics.

I thank Sarah Greene and Helen Kiesselbach Greene for their invaluable help in the preparation of this edition. Information on George Shull was drawn from the archives of Cold Spring Harbor Laboratory and the most informative essay *Mendel in America: Theory and Practice, 1900–1919* by Diane B. Paul and Barbara A. Kimmelman, published in *The American Development of Biology*, edited by Ronald Grainger, Keith Benson, and Jane Maienschein (University of Pennsylvania Press, 1988).

Cold Spring Harbor
September 1998

John R. Inglis

The Structure and Reproduction of Corn

50TH ANNIVERSARY EDITION

The Structure and Reproduction of Corn

T. A. Kiesselbach [1]

OBJECTIVES

The primary purpose of this paper is to summarize investigations concerning the structure and reproduction of the corn plant *(Zea mays* L.*)*. Because of wide interest in the reproductive process in connection with breeding and genetic studies the floral development, fertilization, and embryology are especially stressed. The morphology and development of the vegetative organs is also included. The botanical relationship, origin, genetics and breeding of corn are briefly considered.

When these studies of the corn plant were begun at the Nebraska Agricultural Experiment Station, much along these lines had been done but many points were still unsettled. While this work was in progress many papers appeared which have supplied much of the information that was lacking. The original data herein presented often serve to confirm the results published by others. The literature bearing on the subject matter of this paper is so voluminous that a general review thereof seems inadvisable. Many of the pertinent papers are cited however, some of which include extensive bibliographies and reviews.

DESCRIPTION AND BOTANICAL RELATIONSHIP

Corn as a member of the grass family, *Gramineae,* has many characters common to other grasses, such as conspicuous nodes in the stem, a single leaf at each node, the leaves in two opposite ranks, i.e., distichous, each leaf consisting of a sheath surrounding the stem and an expanded blade connected to the sheath by a blade joint. As in other grasses there is a tendency to form branches at the nodes, and adventitious roots at the base of the internodes. The lower branches may take root and develop into stems known as tillers or suckers which resemble the main stem, whereas the others develop as rudimentary or functional ear shoots. All stems and branches normally form terminal inflorescences in which the flowers are produced in spikelets in which each flower is enclosed between two bracts, the lemma or flowering glume and palet. At the base of each spikelet are the two glumes without flowers, sometimes called empty glumes *(67).* [2]

Corn is normally a monoecious plant (Fig. 1), with its functional staminate flowers borne in the tassels which terminate the stems, and its functional

[1] Acknowledgment is made for the assistance of N. F. Petersen, deceased, with much of the anatomical work herein reported, and to Miss Marie A. Corkle for many of the drawings made either from photomicrographs or with the aid of the camera lucida.

A number of illustrations are reproduced from previous publications of this Station.

[2] Italicized numbers in parentheses refer to "Literature Cited," page 93.

Fig. 1.—Corn in flower. This is a monoecious species in which the pistils and stamens are produced in different flowers of the same plant.

Below, corn field with plants in flowering stage.

Above, ear shoot with receptive silks, and tassel shedding pollen.

pistillate flowers borne in the ears which terminate all but the basal branches or tillers. The heritable vegetative size of plants varies greatly with variety and regional adaptation. Varieties from various parts of the United States have ranged from 4 to 11 feet tall when grown at the Nebraska Agricultural Experiment Station *(43)* and have required from 85 to 155 days to mature.

In most grasses the elongated internodes become hollow, but in members of the corn tribe, *Maydeae* (now sometimes called *Tripsaceae*), and the sorghum tribe, *Andropogoneae,* the stems remain solid. These tribes are evidently closely related and have other characters in common, such as a tendency to have the spikelets in pairs, one sessile, the other pedicellate, each spikelet potentially two flowered.

Contrast of Maydeae and Andropogoneae

In the *Andropogoneae* (sorghum tribe) the sessile spikelet of each pair usually has one perfect flower, whereas the pedicellate one may have only staminate flowers or it may be sterile. The staminate and perfect flowers are thus intermixed throughout the inflorescence.

In the *Maydeae* (corn tribe) the flowers are characteristically either staminate or pistillate and the two kinds are produced either in different inflorescences as in corn, *Zea mays* L., and teosinte, *Euchlaena mexicana* Schrad., or in different parts of the same inflorescence, the pistillate below the staminate as in the species of gama grass, *Tripsacum* spp., the gama grass of the northern states being *Tripsacum dactyloides* L. *(94* and *97).*

ORIGIN OF CORN

NO WILD PLANT is known from which corn could readily have been derived. This might be accounted for by the assumption that the wild corn plant has become extinct. Teosinte is usually regarded as very closely related to corn, not only because of its morphological resemblance, but also because it can be hybridized readily with corn, the progeny being fertile *(63).* Upon crossing, the chromosomes pair normally and crossing-over takes place between the corn and teosinte chromosomes. Teosinte is also susceptible to such common corn diseases as corn smut and corn rust.

Any satisfactory explanation of the origin of corn must also account for its close relationship to teosinte. It has been suggested *(17* and *48)* that corn arose from a cross of teosinte with a grass belonging to the *Andropogoneae* (sorghum tribe). This, however, seems doubtful as it neither has been shown that teosinte can be crossed with any member of the *Andropogoneae* nor that if such a cross were made, the result would be corn.

Following extensive studies of crosses between corn and gama grass, Mangelsdorf and Reeves *(63)* recently suggested that teosinte arose as a natural cross between corn and a species of gama grass. This would explain the close relationship of teosinte and corn, and would eliminate teosinte as an ancestor of corn, but does not account for the origin of corn.

The conclusion reached by Montgomery *(66)* and by Weatherwax *(91)* that both corn and gama grass have arisen from an unknown common ancestor by independent lines of descent is accepted. This, together with the Mangelsdorf and Reeves theory *(63)* of the origin of teosinte by natural hybridization of corn and gama grass, would account for the origin of all three genera of the corn tribe.

Based on the evidence of comparative morphology, it is concluded that modern naked corn has arisen from an assumed primitive ancestor bearing glume-enclosed seeds on the brittle rachis of a terminal, perfect-flowered inflorescence. The active agents involved in the changes from primitive to modern corn are regarded as mutation, hybridization, and selection, whether under natural competitive conditions in the wild or under domestication by man.

There has been much speculation *(85, 62* and *63)* as to whether pod corn *(Zea tunicata)* may be regarded as the wild progenitor or whether it has arisen more recently by mutation as did teopod *(53)*. When pure or homozygous, this plant has no ear and produces seeds only in the tassel. Podded ears are produced only on plants which are heterozygous for the pod character. The glumes of the podded ear cover the seeds as in more primitive grasses, but tend to be abnormally large and often resemble husks rather than ordinary glumes. The glumes of the fertile tassels are also quite different from those of the staminate inflorescences of other *Maydeae*. Glume size is highly variable and appears to be controlled by modifying factors which have arisen by mutation. It is plausible that the dominant gene now responsible for the pod character was present in primitive corn, but this does not signify that pod corn as we know it is or closely resembles the primitive corn.

The geographic point or origin is generally conceded to be somewhere in the tropics *(47* and *63)* of Central or South America, with the latter seeming most probable. This belief is based upon archeologic and ethnologic evidence and upon the theory that the birthplace of a new species is likely to be found in the region of its greatest variability (17 and 87).

The most striking difference between corn and the related grasses is in its pistillate inflorescence which develops into the corn ear. In the other members of the corn tribe such as teosinte and the gama grasses the pedicellate spikelet is rudimentary or absent, causing the seeds to be produced in two single rows on distichous spikes. In corn both spikelets of each pair are practically sessile and each develops a single seed. The seeds are thus arranged in double rows and the spike, instead of bearing two double rows, commonly has four or more double rows, producing eight or more rows of kernels on the rachis or cob (Fig. 2). The sheaths of the leaves of the pistillate branch are much broader than in the ordinary leaves and the internodes of the branch are short so that the ear is completely covered by the leaf sheaths or husks. The outer husks are arranged in two ranks but the inner husks are in several ranks, or in about as many ranks as there are double rows on the ear.

FIG. 2.—Photographs. Superficial views of ear and kernels of mature dent corn.
Left, ear with husks. X 0.3.
Upper right, cross section of ear, showing kernel attachment to cob. X 0.8.
Lower right, anterior, posterior, and lateral surfaces of grain. X 1.
The germinal side of the kernel typically faces the tip of ear, but faces opposite in the case of kernels developed from the lower flower of spikelet.

Distinct progress has been made in recent investigations (1, 2, 22, and 81) as to the homologies of tassel and ear structure and the relation of their development to that of other parts of the plant. The plant is regarded as comprised of structural units (phytomers) whose parts, although modified, are homologous whether in the vegetative or reproductive regions of the plant. The units consist of a section (node and internode) of the axis, with an attached leaf and axillary bud. Their basic distichous arrangement extends throughout the main stem, branches, tassel, and ear, but this has become heritably modified in the inflorescences by such phenomena as condensation (telescoping of successive internodes), multiplication, and change in the number of spikelet pairs per whorl on the rachis.

THE LIFE CYCLE OF CORN

As AN INTRODUCTION to the more detailed study of the vegetative and reproductive structures of the corn plant, its life cycle is briefly summarized before presentation of the experimental data. The entire corn plant, including roots, stems, leaves, tassels, and ear shoots, prior to the time of spore formation by reduction–division in the spore mother cells, is of the sporophytic generation, in which all the cells of typical varieties have nuclei with 20 chromosomes.[3] Ten of these chromosomes are derived from the nucleus of the egg cell and ten from the sperm nucleus which fertilizes the egg.

In the anthers of the tassels each spore mother cell (microsporocyte) divides by reduction-division or meiosis, forming four spores, each with 10 chromosomes. In each spore the nucleus divides, forming a vegetative or tube nucleus and a generative nucleus. The latter divides, forming two sperm cells. Thus the mature pollen grain contains a male gametophyte consisting of three haploid cells.

In the development of the pistillate inflorescence, or ear, the single spore mother cell (megasporocyte) in the one ovule of each functioning flower also forms four spores by meiosis, each with a nucleus having 10 chromosomes. Three of these spores degenerate. By three nuclear divisions without wall formation within the one remaining spore, an eight-nucleate embryo sac is formed with four nuclei at each end. Two of these, one from each set of four and known as polar nuclei, approach each other and lie in close contact preliminary to being included in the primary endosperm nucleus.

The cytoplasm becomes organized around each of the three nuclei at the chalazal end of the embryo sac, forming a group of three antipodal cells. These soon begin to divide until approximately 30 or more antipodal cells are formed. Cells are also organized around the nuclei at the micropylar end of the embryo sac. One of these enlarges and becomes the egg cell while the others form the adjacent synergids. This group of three cells is often called the egg apparatus.

The mature embryo sac or female gametophyte, when ready for fertilization, contains the three-celled egg apparatus, the group of antipodal cells at the opposite end of the sac, and the two large polar nuclei located centrally above the egg.

When the silks of the ear shoot are pollinated, the pollen grains germinate and send pollen tubes into and toward the base of the individual silks. Usually the tube of only one pollen grain reaches the micropyle where it grows between the cells of the nucellus until it enters the embryo sac, ruptures, and releases the two sperms. The nucleus of one sperm fuses with that of the egg to form the zygote, restoring the diploid or $2n$ number of 20

[3] These are the numbers most commonly found in corn, variations from which will be discussed later. The practice will be followed throughout the paper of giving what is typical or usual and not including the numerous variations which are known to occur.

chromosomes in the cell nucleus. This number persists thereafter in the somatic cells of the sporophyte, all of which are formed by mitotic division.

The other sperm fuses with one of the two polar nuclei and this in turn fuses with the other polar nucleus, forming the primary endosperm nucleus. This nucleus has 30 chromosomes, 10 derived from each polar nucleus and 10 from the sperm nucleus. The process in which there is one fusion to begin the embryo and another to begin the endosperm is called double fertilization. Although double fertilization was first discovered at the close of the last century *(31)*, it has now been found in so many plants that it may be safely assumed to be almost universal among the flowering plants. However, in some genera the endosperm is largely consumed by the developing embryo and only minute traces remain in the mature seed.

In the lower groups of plants all parts of the plant are either gametophytic with $1n$ chromosomes in each nucleus, or sporophytic with $2n$ chromosomes in each nucleus. The $3n$ endosperm of corn is thus neither typically sporophytic nor gametophytic. This has led to a great deal of speculation as to whether it is of the gametophytic or sporophytic generation. Some workers have considered the endosperm as an abnormal twin of the embryo, whereas others regard it merely as modified tissue of the gametophyte whose continued development is dependent upon the stimulus contributed by the sperm nucleus. This latter view appears the more tenable. Whatever the explanation, the endosperm is a modified tissue that serves for the nutrition of the developing embryo or seedling, or both as in corn.

The embryo sac following double fertilization has all three types of cells with respect to number of chromosomes: (1) the antipodals are haploid, with $1n$ (10) chromosomes; (2) the embryo (beginning with the zygote) is diploid, with $2n$ (20) chromosomes; and (3) the endosperm is triploid, with $3n$ (30) chromosomes.

The synergids soon disappear. The group of antipodals tends to persist for some time, often being found in the mature kernel.[4] The fertilized egg develops into the embryo. The primary endosperm nucleus, after a period of free nuclear division in the cytoplasm of the embryo sac, forms cells that develop into the endosperm which comprises the bulk of the mature kernel.

The kernel is thus found to consist of (1) parts of the mother sporophyte, namely the pericarp, remnants of the nucellus and seed coats, and the pedicel, (2) the new sporophyte or embryo, (3) the endosperm, and (4) sometimes the small group of antipodals or remnants of the original $1n$ female gametophyte.

The embryo consists of a short stem to which are attached two modified leaves (the scutellum and coleoptile), several typical foliage leaves, a primary seminal root, and generally several lateral seminal root initials. During germination the embryo develops into a seedling, while the food stored in the

[4] This tendency in corn and perhaps some of its relatives is unusual among the grasses, the antipodals disorganizing about as soon as the endosperm becomes well established in most grasses.

endosperm is digested and absorbed by the scutellum.

The seedling develops additional leaves which together with those present in the seed grow into mature leaves. The small stem develops into the stalk, which attains full size at the time of tasseling. Adventitious crown roots which make up the major part of the root system of the mature plant form at the base of each of the lower internodes of the stem.

While the seedling is still small, the tassel begins to develop and by midsummer attains full size and bears staminate flowers which produce the pollen. The tassel withers when the pollen has been shed. Branches of the stem form at about the same time as tassel formation begins. Several branches at the base may become tillers but most of them develop into ear shoots. Most of the lower ear shoots degenerate, but those which persist bear functional flowers, each of which produces an ovary with a single ovule. The ovary ends in a long style or silk which is stigmatic to within about an inch of its base. This serves to catch the pollen when shed by the tassels. A pollen tube with two sperms grows from the pollen grain to the embryo sac within the ovule where double fertilization takes place, and the next sporophytic generation is thus started.

DEVELOPMENT AND STRUCTURE OF VEGETATIVE PARTS

As AN AID to tracing the origin of the vegetative parts of the corn plant, this anatomical discussion is prefaced with a brief description of the kernel and its parts. Their development and origin are discussed later under other topics. It would seem to be a logical sequence to examine the structure of the kernel, followed by a study of the germinating seed, the seedling, and the growing plant.

The Kernel or "Seed"

The corn kernel (Fig. 3) is not merely a seed but a one-seeded fruit, in which the seed, consisting of embryo and endosperm and remnants of the seed coats and nucellus, is permanently enclosed in the adhering pericarp. The structure of the mature kernel, also known as a caryopsis, is here described. Upon shelling, the flower stalk or pedicel commonly remains attached to the base of the kernel.

PERICARP

The pericarp or transformed ovary wall forms the tough outer covering of the kernel, and furnishes protection for the interior parts. It thus takes the place of the seed coats or integuments of ordinary seeds, which in corn are reduced to inconspicuous remnants in the mature kernel. A thin, suberized nucellar membrane derived from the outer epidermal wall of the nucellus persists as a continuous covering between the aleurone and the pericarp *(71)*. At the apical end of the kernel the pericarp bears the silk scar, and at the basal end it merges into the tissues of the pedicel or tip cap as it is often called.

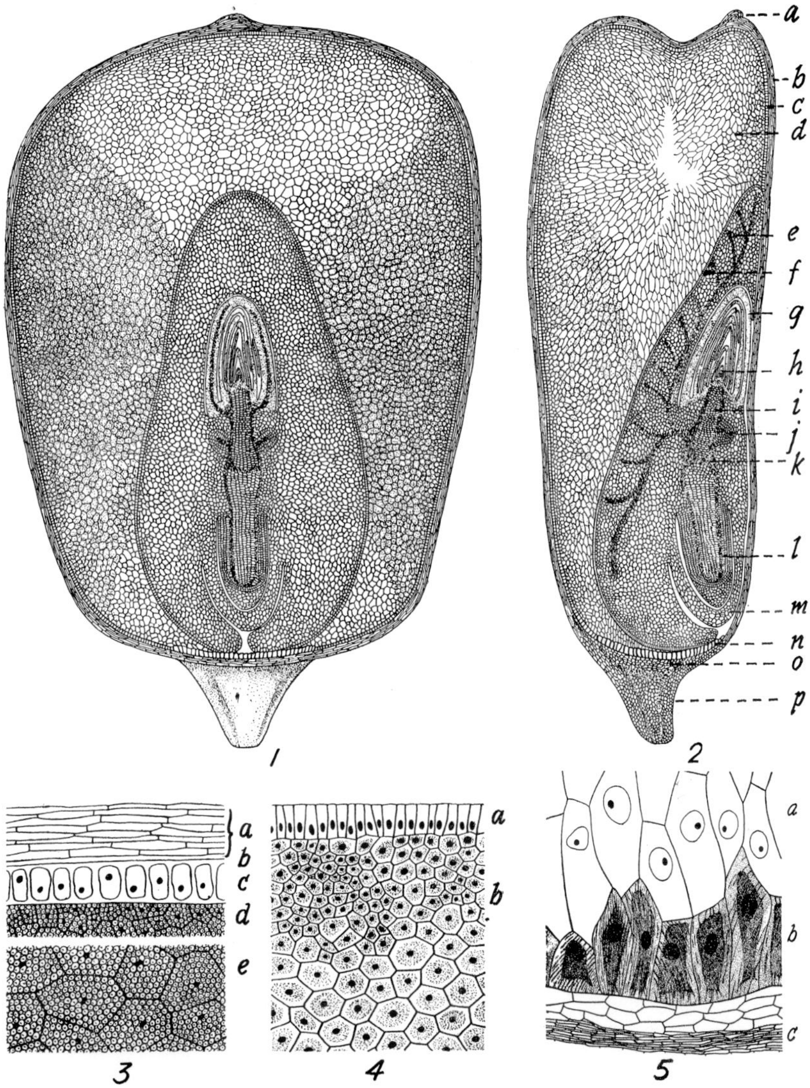

FIG. 3.—The mature kernel.

1 and 2, vertical sections in two planes of a mature kernel of dent corn, showing arrangement of organs and tissues. a, silk scar; b, pericarp; c, aleurone; d, endosperm; e, scutellum; f, glandular layer of scutellum; g, coleoptile; h, plumule with stem and leaves; i, first internode; j, lateral seminal root; k, scutellar node; l, primary root; m, coleorhiza; n, basal conducting cells of endosperm; o, brown abscission layer; p, pedicel or flower stalk. X 7.

3, enlarged section through pericarp and endosperm. a, pericarp; b, nucellar membrane; c, aleurone; d, marginal cells of endosperm; e, interior cells of endosperm. X 70.

4, enlarged section of scutellum. a, glandular layer; b, interior cells. X 70.

5, vertical section of the basal region of endosperm. a, ordinary endosperm cells; b, thick-walled conducting cells of endosperm; c, abscission layer. X 350.

ENDOSPERM

When the pericarp and adhering remnants of the seed coats and nucellus are removed, which can easily be done after soaking the kernels in water, only the endosperm and embryo are left. The endosperm makes up the greater part, the relatively small embryo being situated on one side near its base. Except for the outer (surface) layer, the endosperm consists of cells filled mainly with starch grains. At the base of the endosperm, opposite the region of placentation, the cells of its surface layer are modified as conducting cells (Figs. 3–5, and 57) and apparently serve for conducting food from the mother plant to the growing endosperm and indirectly to the embryo *(96)*. The remaining surface cells form the normal aleurone layer, the cells of which contain aleurone grains and oil, but no starch.

EMBRYO

The embryo or young corn plant is embedded near one face of the endosperm at the base of the kernel or caryopsis. It has a central axis (Fig. 3–1 and 2, Fig. 4) which is terminated at the basal end by the primary root and at the other end by the stem tip. The stem comprises five or six short internodes and bears a leaf at each node. The first leaf, known as the scutellum, is attached to the first or scutellar node. It never functions as a true foliage leaf but it is modified as a food storage organ and serves to digest and absorb the endosperm during growth of the embryo and seedling *(28, 33,* and *76)*. Its outer layer where in contact with the endosperm is specialized to produce enzymes for digesting the starch in the endosperm (Fig. 5).

The second leaf, the coleoptile, is attached to the second or coleoptilar node and is modified as a protective covering for the plumule or first bud

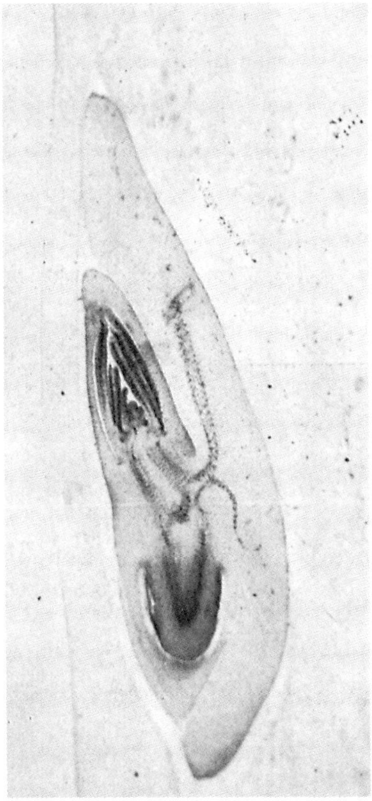

FIG. 4.—Photomicrograph. Vertical section of a mature embryo within the dormant kernel, showing the scutellum at right, the primary root enclosed in the coleorhiza below, and the plumule with its central growing point and leaves within the coleoptile above. X 10.

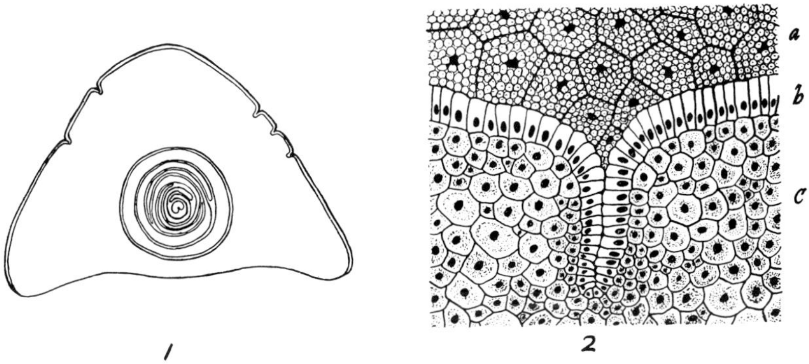

FIG. 5.—1, cross section of scutellum showing region where glands are usually most abundant. X 10.

2, section through scutellum showing gland. a, endosperm; b, epithelium on posterior side of scutellum, with gland; c, ordinary tissue of scutellum. X 110.

of the plant which it spearheads through the soil upon germination. The plumule consists of that part of the stem which extends above the coleoptilar node, and four or five small foliage leaves each rolled up inside those below it, forming a cone within the coleoptile.

The first internode of the stem between the scutellar and coleoptilar nodes lengthens quickly during germination and serves to elevate the coleoptile to the soil surface.

The primary root is enclosed by the coleorhiza or root sheath, and usually the initials of two or more adventitious seminal roots arise at the base of the first internode of the stem. Both the first and second internodes *(7)* are transitional from stem-like to root-like structure.

PEDICEL

The tissues of the pedicel or flower stalk merge into those of the ovary wall without close demarcation between the two. Upon shelling from the cob, the pedicel with adhering lemma and palet usually remains attached to the kernel (Figs. 2 and 3). At the time of ripening, however, before the tissues have toughened by desiccation, the kernel often breaks off in the plane of seed attachment exposing the outer surface of a thin, dark brown abscission layer that corresponds to the hilum of funiculate species. This layer, also regarded as a closing, protective tissue, forms a plane of cleavage that extends entirely across the "hilar orifice" *(36)* which is co-extensive with the chalaza, where the nucellus and seed coats join. Its location is seen in Figures 3, 57, and 62–4, just below the base of the endosperm.

Seed Germination and Seedling Development

When the corn kernel is placed under moisture and temperature conditions favorable for germination, growth activity is quickly resumed by the embryo. In official seed testing laboratories a period of five or six days is considered ample to determine viability at the optimum temperature of 30° C. (86° F). Planted in the field during the usual time in the spring, emergence of the seedling occurs in about seven to ten days, varying with soil temperature and moisture. After the soil has warmed fully, emergence may take place in approximately five days. In Figure 6 is shown the development after 35 and 55 hours of germination under favorable conditions.

Healthy, well preserved seed usually has high viability, 95 to 100 per cent being common. The chief cause of low germination is injury from freezing. This results from exposure to low temperatures while the moisture content is still high. Seed containing 14 per cent or less of moisture will endure any amount of freezing to which it is likely to be exposed, without injury *(46)*. A second cause of reduced viability is infection with seed-borne or soil-borne rot organisms which either prevent germination or destroy the seedling at a very early stage. Seed treatment with commercial disinfectants or protectants, as Arasan and Spergon, has been reported from many states to

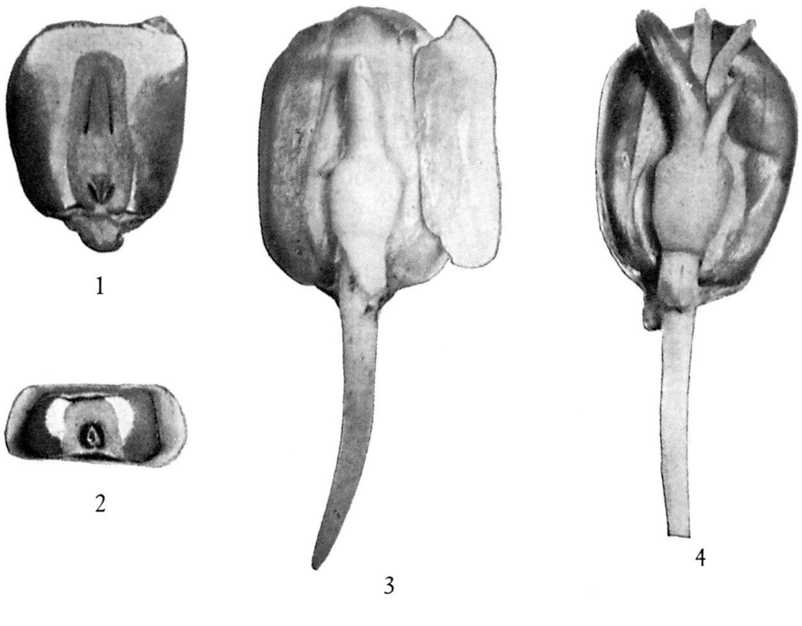

FIG. 6.—Photographs.

1, vertical section of mature kernel of dent corn. X 3.

2, cross section of mature kernel of dent corn, through the primary root. X 3.

3, growth after 35 hours favorable germination (pericarp laid back). X 4.

4, growth after 55 hours germination, showing lateral seminal roots arising just above scutellar node. X 4.

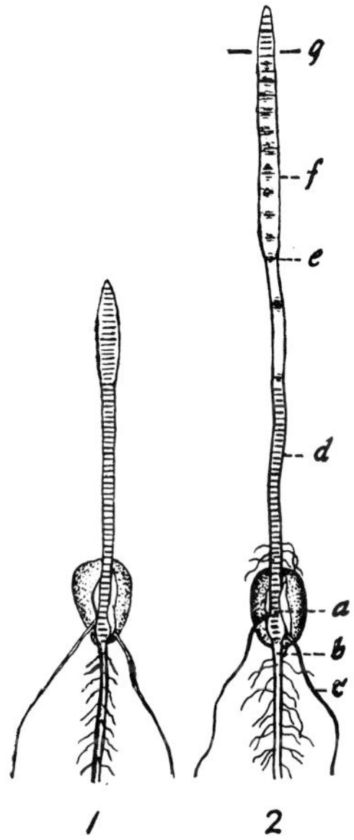

FIG. 7.—Germinated seed, showing region of cell division and elongation at upper end of first internode. X 0.6. In case of deeply planted seed this intercalary growth serves to lift the tip of coleoptile to the soil surface, which facilitates emergence of the enclosed growing leaves and stem tip of the plumule. The coleoptile itself shows rather uniform growth throughout most of its length by cell enlargement.

1, first node and internode and coleoptile of young growing sprout marked at 1 mm. intervals with India ink.

2, same sprout 32 hours later. The wider spaces indicate the region of elongation. Cell division is confined to a very short meristematic region of about ½ millimeter length just below the coleoptilar node. a, scutellar node; b, primary root; c, lateral seminal root; d, first internode of stem; e, coleoptilar node; f, coleoptile; g, soil surface.

increase the seedling emergence by about 5 to 20 per cent. Improper storage conditions which delay drying may cause damage with loss of viability.

Under favorable storage conditions, seed corn may retain its viability for several years if properly matured and dried to a safe moisture content *(43)*. It can be used for field planting up to four years of age without serious stand reduction. After six years the germination usually begins to drop rapidly, although some of the seeds may remain viable twice that long.

During germination the organs of the embryo which were formed during the development of the kernel and have remained dormant in the dry seed, resume their growth and development. The primary root and its enclosing sheath, the coleorhiza, elongate and break through the pericarp, the root soon breaking through the end of the coleorhiza. The plumule and its enclosing sheath, the coleoptile, begin to elongate and also break through the pericarp of the kernel. At first the coleoptile grows faster than the plumule but when it reaches the surface of the soil and is thus exposed to the light, it soon ceases to grow and the plumule breaks out through its tip. If the seed has been planted deeply the first internode, or the part of the stem between the scutellar and coleoptilar nodes, also elongates by intercalary growth at its upper end (Fig. 7), thus helping to raise the coleoptile to the surface. About the time that the tip of the coleoptile reaches the soil surface the first crown roots appear immediately above the coleoptilar node (Fig. 8) and later an additional whorl of roots forms at the base

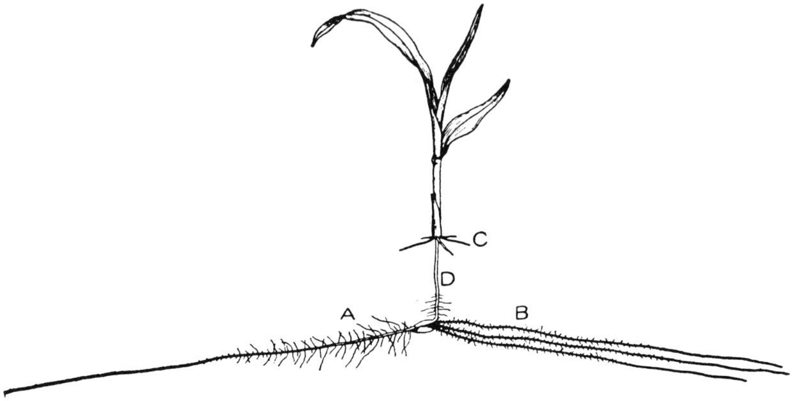

FIG. 8.—Corn seedling two weeks after planting, showing the seminal and crown roots in normal position. A and B, primary and lateral seminal roots, respectively; C, initial whorl of adventitious crown roots; D, first internode of stem, which serves to elevate the tip of coleoptile with enclosed plumule to surface of ground. The seedling leaves were already well organized in the dormant embryo. X 0.3.

of each succeeding 6 to 10 internodes of the stem. These crown roots soon form the major part of the root system of the plant.

As the stem elongates, its growing point continues to form additional leaf initials, one at each node (Fig. 9), which later develop into leaves. When all the leaves are started but before many of them are unrolled, the growing point of the stem elongates, becomes branched, and starts to differentiate (Fig. 10) into the young tassel. Before the tassel has differentiated, while the growing point of the stem is still giving rise to additional leaf initials, branch buds (Fig. 9) have formed in some of the lowermost leaf axils. Those in some of the lower axils may develop into tillers, or in non-tillering corn they soon die and disintegrate. The rest will form ear shoots, most of which soon degenerate with only the upper one or two developing functional ears. The number of tillers, degree of development of rudimentary ear shoots, and number of ears developed depend partly on inherited tendencies of the corn and partly on environmental conditions. Up to the stage of tassel differentiation, which is reached in approximately three weeks, the young plant may be thought of as a seedling. Initials have then been laid down for all the nodes and internodes of the stem, all the leaves, tillers, ear branches, and the primary and adventitious root systems. Martin and Hershey *(64)* observed that 90 per cent of the final number of vascular bundles had been laid down in the lower internodes at the time of tassel differentiation. From this point in the discussion the development of the various parts will be considered individually.

The Root System

The root system of corn, as of other grasses, consists of two sets of roots: (1) seminal roots whose initials are present in the embryo, and (2) adven-

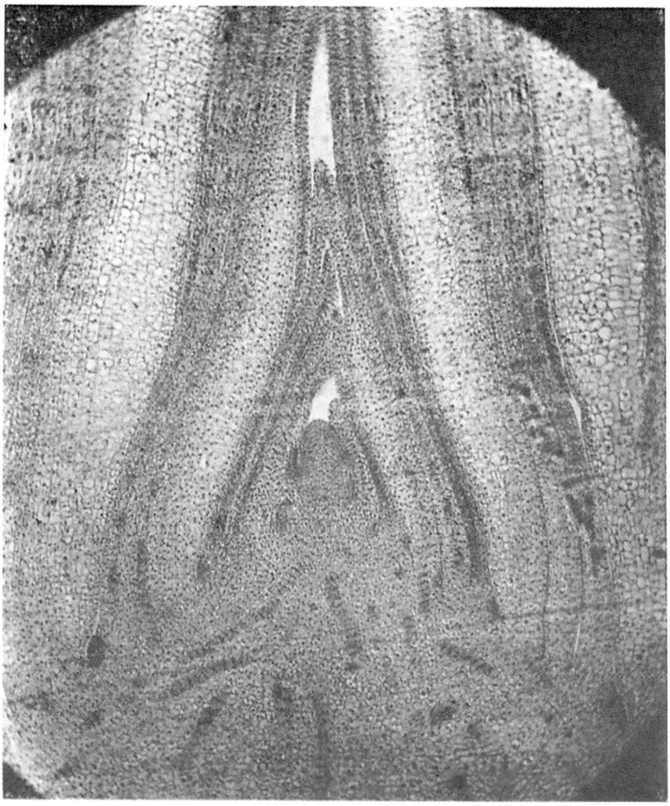

Fig. 9.—Photomicrograph. Vertical section through portion of corn seedling above coleoptilar node, seven days after planting, showing within the coleoptile: the central hemispherical growing point of stem surrounded by about eight leaves laid down by growing point, and branch buds (tiller or ear) in several leaf axils. Initials of the vascular system show below. X 20.

titious roots which arise from stem tissue after germination (Figs. 8, 11, 12, and 13). These are often called *temporary* and *permanent,* respectively; but it has been shown repeatedly that the seminal roots may persist and function throughout the life of the plant.

SEMINAL ROOTS

The seminal roots consist of the radicle or primary root and a variable number of lateral roots which arise adventitiously at the base of the first internode of the stem, just above the scutellar node. The radicle is always present except when killed by some injury such as freezing. It may be the only seminal root as is frequently the case in some flint corn varieties, or there may be averages of one to seven lateral seminal roots in various

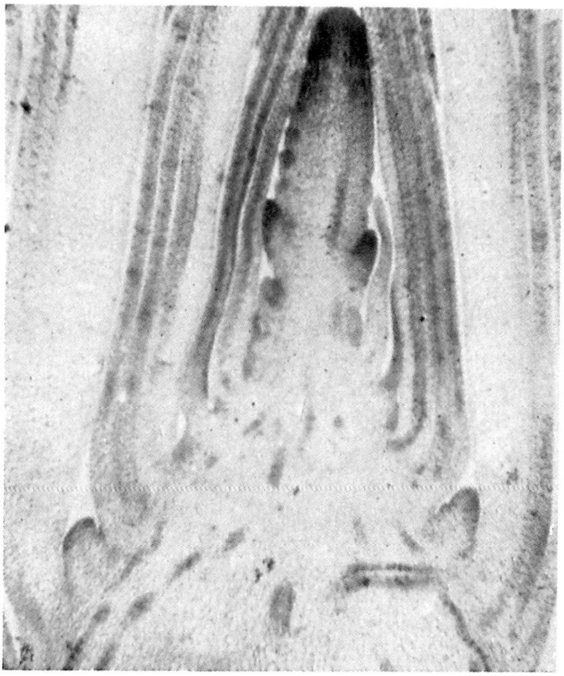

FIG. 10.—Photomicrograph. Vertical section through portion of corn seedling 21 days after planting, showing: differentiated tassel surmounting the stem; the two uppermost ear-shoot buds, each with first husk initial; and the final number of about seven leaves above the uppermost ear shoot. X 16.

varieties of dent corn. It is rather common to include in addition to these a number of minor laterals that may form adventitiously from the pericycle elsewhere in the first internode of the stem. Wiggans *(100)* found the number of seminal roots to vary from 1 to 13 per plant.

Under ordinary field conditions at the usual time of planting corn, the seminal roots grow in a nearly horizontal direction for some distance before turning downward *(99)*. They form such a small part of the total root system of the plant that they are of greatest importance during the early growth of the seedling before the adventitious crown roots of higher internodes have become well established. Replicated field tests of Krug corn from which the seminal root system was severed at the age of three weeks by means of pulling wire snares which had been inserted about the first internode at the time of planting germinated seeds, showed a loss of 9 per cent in yield of grain.

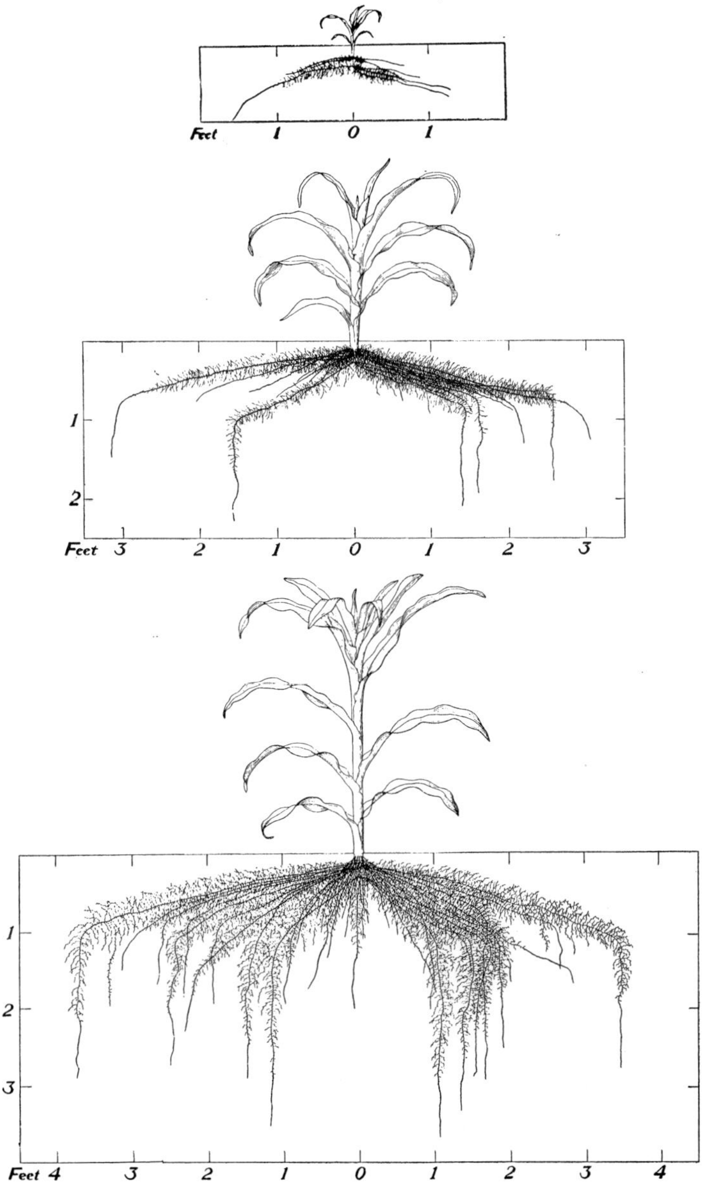

FIG. 11.—Stalk and root system of a typical corn plant (Krug variety) two, four and six weeks after planting in 1933 on the Experiment Station farm at Lincoln. The seminal and crown roots may be traced separately. About 25 per cent of the fine branch roots are included.

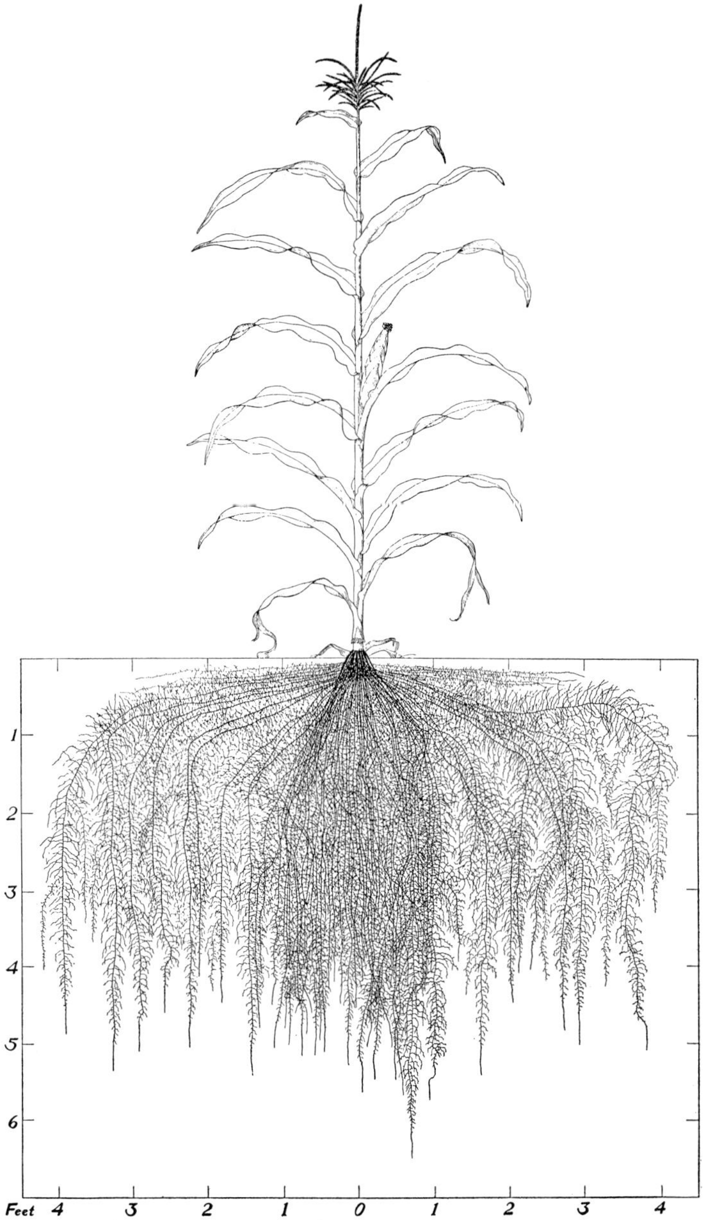

FIG. 12.—Mature stalk and root system of a typical corn plant (Krug variety) grown on the Experiment Station farm in 1933. About 25 per cent of the fine branch roots are included. In both Figures 11 and 12 the roots were drawn in position during excavation, except that they are shown in a single plane.

ADVENTITIOUS ROOTS

The term *adventitious* roots has come to apply to all of those roots that arise in the basal intercalary meristem of the lower internodes of the stem. Through common usage this is synonymous with *crown* roots, so designated because of origin in the basal portion of the growing stem known as the crown. Following the seedling stage, they constitute the principal part of the root system. Any aerial brace roots are included in this category.

The first whorl of four or five crown roots appears at the base of the second internode about as soon as the tip of the coleoptile reaches the soil surface. A few of the succeeding higher internodes may have approximately the same number, after which the successive ones have more and larger roots up to a little above the soil surface. Those from about the lower five internodes, like the seminal roots grow horizontally for some distance before turning downward, while those from the higher internodes which appear later in the season (Figs. 11, 12) grow downward at once. Hays *(34)* suggested that this peculiar behavior of the earlier roots of corn might be due to the low temperature of the deeper soil at the time they are forming. The change in direction of growth occurs after about four weeks, when the deeper soil has become warm. If the season is dry and warm, there might thus be less horizontal growth than if it is wet and cold. With corn planted in midsummer there was little or no horizontal growth, all roots starting downward at once. Due to this horizontal growth of the earlier roots they may spread over an area about 8 feet in diameter, while the later roots may be confined to an area of but 12 to 18 inches in diameter, just under the hill or stalk.

Fig. 13.—Base of main stalk and attached growing tiller and seminal roots. The crown roots have been cut off to show their arrangement at the crown.

In locally adapted varieties of corn at Lincoln, Nebraska, Weihing *(99)* found that an average of nine internodes bear functional roots (Fig. 13), eight of these below the surface of the soil, and one just above the surface. Except for a few of the lower internodes, in which the numbers of roots were about equal, there was a gradual increase upward. The functional crown roots averaged 85 per stalk, with a total combined length of about 350 feet. These roots are all branched and rebranched so that the total length of the branches greatly exceeds that of the main roots. Weihing estimated the total length of all the roots at about 6 miles per plant under eastern Nebraska conditions.

In addition to the functional roots of the crown, three or four higher aerial internodes may bear nonfunctional roots which fail to reach the ground. Under field conditions of rank growth, some or all of these may become functional and are then known as brace roots.

FORMATION AND GROWTH OF ROOTS

Initials of root tissues.—The end of each root is covered by a rootcap (Fig. 14–1), back of which are the three initials of the real root tissues, each consisting of a group of cells which do not themselves become differentiated, but by dividing form cells which by further division and differentiation form the cells of the root tissue.

One of these groups is the calyptrogen which is the inner layer of the rootcap and divides successively to form additional layers of cells for the rootcap. Just back of this is the small single-layered periblem-dermatogen group which is the common initial of the epidermis and the cortex. These cells divide tangentially, i.e., parallel to the root surface, to form two layers of cells *(102)*. The outer of these acts as a dermatogen, forming cells which differentiate to form the epidermis. The cells of the inner layer divide to form several layers of cells as well as to increase the number of cells in each of these layers. After some time only the innermost layer continues to divide tangentially, the result being that there are regular radial rows of cells in the inner part of the periblem while the cells are arranged irregularly in the outer layers. This inner layer finally differentiates as the endodermis, while the rest of the periblem forms the parenchyma of the cortex.

The third group, which is the initial of the plerome or central part of the root, is just over the common origin of the dermatogen and periblem. The stele, consisting of xylem and phloem strands with their surrounding ground tissue, will differentiate in the plerome. Its outermost layer of cells differentiates to form the pericycle, just inside the endodermis.

Differentiation of root tissues.—There is little differentiation of cells in the rootcap, aside from increase in size. Intercellular spaces develop and the surface cells are worn away during soil penetration. The rootcap forms none of the permanent tissues of the root and serves only as protection for the underlying meristematic tissues and as "lubrication" for the root tip as it is pushed through the soil.

The main differentiation of the dermatogen, aside from thickening of the outer walls, is the formation of root hairs (Fig. 14–3). These start as a bulging of the outer wall of an epidermal cell which becomes a long and

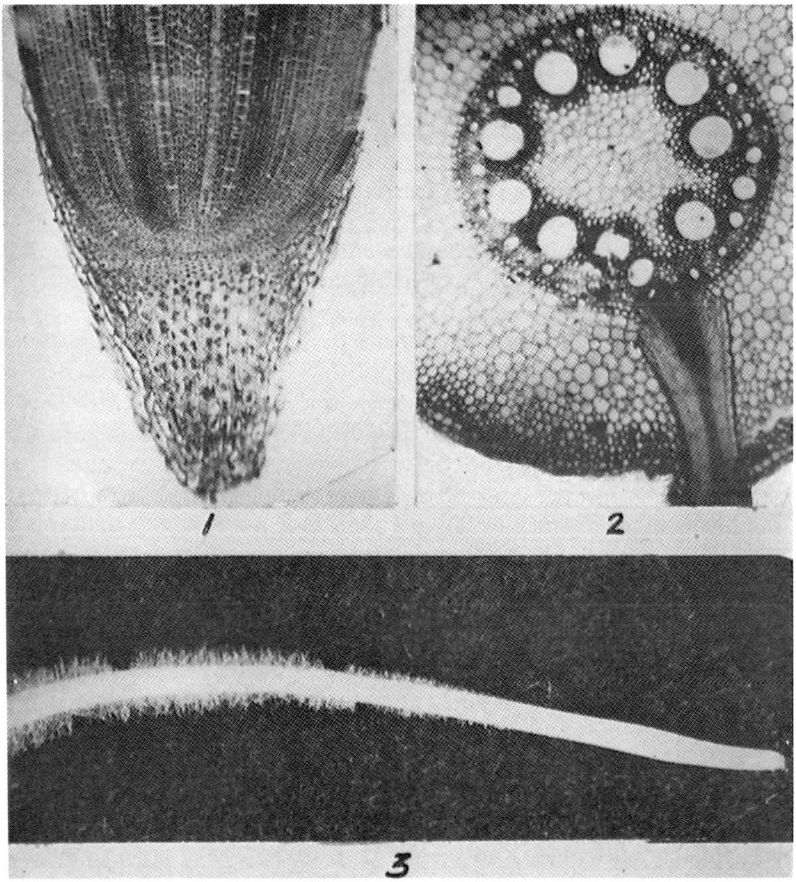

FIG. 14.—Photomicrographs of corn root.

1, longitudinal section of a root tip, showing the rootcap and the developing tissues of the stele. X 50.

2, cross section of a root some distance back of the tip, showing one branch. The stele is shown centrally with an inner circle of large vessels and an outer series of small vessels. Lateral roots originate in the outer layer or pericycle of the stele and grow through the cortex. X 50.

3, surface view of growing root, showing root-hair zone back of tip. X 2.

tubular outgrowth. These hairs greatly increase the root surface and are the main organs for absorbing water and nutrient salts from the soil. The root hairs soon break down and the epidermal cells die and are lost, cortical cells then forming the outer surface of the root. The region of root hairs is the absorbing region of the root. The older part of the root from which the hairs have disappeared becomes mainly a water-conducting organ. Numerous root branches form behind the root-hair zone, each like the main root but smaller and with its own zone of root hairs.

The surface of each root, beginning at the tip, thus shows (1) a rootcap, followed by (2) a zone of rapid cell multiplication which may be but a millimeter long and covered by the rootcap, (3) a smooth region of rapid elongation without root hairs, (4) a zone in which the root is covered by root hairs, and (5) finally a zone where the root hairs and epidermis have broken down leaving the cortical cells as the outer surface of the root. These form a lignified and suberized exodermis.

Most of the cells of the endodermis become conspicuous by having thick walls on their inner surface and remaining thin walled on the outer surface next to the rest of the cortex. The division walls between the cells of the endodermis are thick both at their ends and sides where they join the inner thick wall, and thin where they join the thin outer walls, thus appearing wedge shaped in either cross or longitudinal sections. In addition these walls have a narrow strip of waxy substance surrounding each cell. These are called Casparian strips or, as they may appear as dots in sections, Casparian dots. A few of the cells of the endodermis may remain relatively thin walled, forming so-called transfusion tissue through which water and foods may pass. The peculiar structure of this layer which forms a kind of tube around the stele is supposed to have something to do with getting water into the center of the root and keeping it there, but how it functions is not well established *(102)*.

The outer layer of the plerome, next to the endodermis, becomes differentiated as the pericycle. Inside this, strands of vascular tissue develop, strands of xylem alternating wtih strands of phloem. The number of these varies with the diameter of the roots.

In each xylem strand differentiation of vessels begins just inside the pericycle and advances toward the center of the root. Each xylem vessel is formed from a row of cells, joined end to end, the cross walls having disappeared early in the differentiation of the vessel or trachea. The protoplasm has disappeared in the fully formed vessel, leaving only the lateral cell walls. The first-formed vessels, protoxylem, are smallest and the later-formed vessels, metaxylem, are the largest in diameter and are nearest the center of the root. In some small roots there may be but a single large central vessel although there are several strands of protoxylem. The cells surrounding the xylem and phloem become sclerenchymatous.

The first-formed phloem, or protophloem, is also formed just inside the pericycle. Later additional phloem forms toward the center of the root. The phloem consists of sieve tubes, each formed by the union of several cells end to end, each unit separated from the next by sieve-like end walls. Each of the original cells of the sieve tube is accompanied by a companion cell. In a cross section the sieve tubes are relatively large and rounded while the much smaller companion cells appear as small squares. As the sieve tubes are rather long only a few will show the characteristic sieve plates. Just as in the xylem, the strand becomes surrounded by elongated fibrous cells.

The space between the xylem and phloem strands is filled up with short, thin-walled parenchyma cells or pith. In large roots there is a large central mass of pith, which may be lacking in very small roots.

Origin of root branches.—Root branches come at irregular intervals on the main roots. When a root branches, the initial of each branch is formed from cells of the pericycle (Fig. 14–2). By cell divisions a bulging hemispherical mass of meristematic tissue is formed. The outer part of this forms the beginning of the rootcap, while the cells of the growing point are near the flat basal part. During elongation the new branch is forced through the adjoining tissues of the cortex of the originating root and appears at its surface. As the xylem and phloem tissues are differentiated in the root branch they become joined to those of the older root.

The Stem

The stems of typical eastern Nebraska corn plants consist of about 24 alternating nodes and internodes. The number may vary somewhat according to regional type and environmental effect. Approximately 8 internodes remain very short and under ground, forming an inverted cone-shaped basal end of the stem known as the crown. The crown internodes give rise to the adventitious crown-root system, and adventitious brace roots may develop at the base of one to several aerial internodes. Under favorable growing conditions, the above-ground internodes are distributed over a stem length of 100 inches or more. With its greatest diameter of about 1¼ inches near the ground level the stem gradually tapers toward its top. Each of the lower internodes adjacent to an ear shoot becomes permanently grooved throughout its length as a result of the pressure of the developing shoot during early growth.

DIFFERENTIATION AND GROWTH OF PARTS

The stem tip becomes evident very early in the development of the embryo as a hemispherical mass of meristematic tissue. The initials of the different regions of the stem are not as distinct as in root tips. The dermatogen is distinguished early in stem development but there is no clear distinction between periblem and plerome.

The stem tip differs from a root tip in several ways. The stem tip has nothing corresponding to the rootcap. Leaf initials which form very near the tip of the stem are lacking in roots. In stems both leaf and branch initials are formed superficially from the epidermis and outer part of the cortex, while root branch initials form internally from the outer part of the plerome, breaking through the cortex and epidermis of the main root.

Soon after meristematic tissues are formed the beginnings of the vascular bundles appear, extending through both nodes and internodes. In the nodes all the longitudinal bundles become united by small cross bundles which form the nodal network. From each node bundles extend into the leaf arising from that node. When buds for branches form at the nodes their vascular bundles also become united to the network of bundles in the node. The internodes are very short at first (Fig. 15–1) but begin to elongate (Fig. 15–2) through the formation of new cells by an intercalary meristem that is located near the base of each internode except the first. In the first internode this region of new cell formation occurs at the upper end, as illustrated in Figure

1 2

FIG. 15.—Photographs. Typical stem of young corn plant.

1, lower portion of plant, 32 days after planting, split lengthwise to show development of stem and attachment of leaves. X 0.2.

2, external view of young stem with leaf sheaths removed, 35 days after planting. X 0.5. Stem height 8 inches, tassel length 0.5 inch, plant height 32 inches. This stalk has 17 nodes above ground, each of the lower 9 having an ear branch. Potential root initials occur in the meristematic tissue just above each of these lower nodes. Normally only the upper one or two ears reach the fertilization stage as is shown in Figure 36.

7. The cells later elongate until the internodes make up most of the length of the stem. It is largely this elongation which may take place in several internodes at the same time that causes the rapid growth in height of the corn plant in midsummer. The region of new cell formation in each internode is very short—less than half a millimeter. Cell elongation, however, extends over several millimeters.

When the stalk is erect, growth is equal on all sides; but if the stalk leans or lies flat on the ground, as after storms, it is more rapid on the lower side, thus tending to raise the stalk upright. Even after growth of the internodes has apparently ceased, if the stalk is placed horizontally growth is resumed and is most rapid on the lower side. A wedge of tissue is formed and the stalk again becomes erect. This newly formed tissue will at first appear white in contrast with the green color of the rest of the internode. When, however, the lower part of the stalk becomes too old, such growth will take place only in the upper internodes, and only this part becomes erect.

ARRANGEMENT OF TISSUES

Internodes.—All of the internodes have a similar structure except the first two which are transitional between stem and root. In typical internodes the stem consists of a pithy interior comprised mainly of parenchyma with vascular bundles distributed throughout, and a hard outer shell of epidermis and sclerenchyma. In cross section (Fig. 16) the epidermis is the outer single layer of cells. The walls of these cells are much thickened, lignified, and silicified. Stomata are present but are less abundant than in the epidermis of leaves.

Just beneath the epidermis is a layer of sclerenchyma which varies from one to about ten cells in thickness. This does not form a continuous covering under the epidermis but forms longitudinal bands or strips over the peripheral vascular bundles, alternating with strips of parenchyma beneath the rows of stomata.

Inside this sclerenchyma are several rows of vascular bundles, each surrounded by a large amount of sclerenchyma. These peripheral bundles are smaller and spaced much closer than those in the central part of the internode and are separated from each other and from the peripheral sclerenchyma by narrow tracts of parenchyma.

The strength of the internode depends largely on the amount and degree of lignification of the sclerenchyma next to the epidermis and around the peripheral bundles. These differences are hereditary and are associated with the breaking strength and lodge resistance of the stalks *(4* and *61)*. Various inbred lines and hybrids differ greatly in these respects.

In the development of the larger bundles (Fig. 17–1) the first xylem vessels to differentiate (the protoxylem consisting of ringed and spiral vessels) are formed in the side of the bundle toward the center of the stem. These are

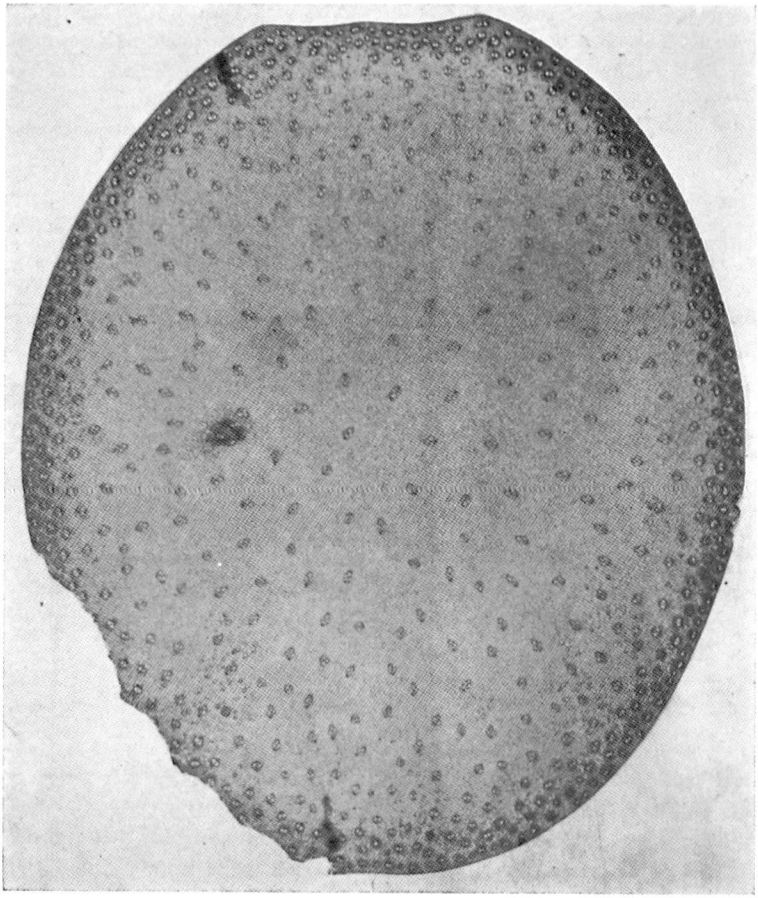

FIG. 16.—Photomicrograph. Cross section of full-grown stem through internode near base of stalk, showing distribution of vascular bundles. X 5.

followed by very large pitted vessels of which there are generally two in each bundle. A group of tracheids forms between them.

At the opposite side of the bundle the first phloem, the protophloem, starts to differentiate, the differentiation progressing toward the xylem so that the phloem, consisting of sieve tubes and companion cells, extends partly between the large pitted xylem vessels. Some of the first-formed ringed vessels in the xylem are broken down by growth of surrounding tissues, leaving a cavity or lacuna. Part of the first-formed phloem also becomes disorganized and compressed but no cavity is formed.

Each of these bundles is surrounded by a bundle sheath, the cells of which become lignified and sclerenchymatous. This sheath is more massive opposite the phloem and xylem of the bundle. These masses of sclerenchyma are sometimes designated as the phloem and xylem caps.

Nodes.—The structure of the nodes is more complicated than that of the internodes (Fig. 17) owing to branching and cross connections between the bundles forming the nodal complex. The bundles with their cross connections fill up most of the space, so there is relatively little parenchyma. In the zone of leaf-sheath attachment at the node, the epidermis of the adjacent internodes becomes continuous over the leaf. In the nodes, where no great elongation takes place, no protoxylem lacunae are formed. The arrangement of the tissues in the bundles is more irregular and no bundle caps are formed.

As the bundles enter a node from below, they enlarge in diameter and form cross connections, creating a tangled network which extends for several millimeters along the length of the stalk. Many of them branch in passing through the node, and some even though not branching turn abruptly sidewise for a short distance. This turning may be due in part at least to unequal growth in diameter of the younger, higher internode and the older, lower internode. Some bundles pass from the node into the base of the leaf sheath. These may be the small peripheral bundles or branches from the larger bundles deeper in the stalk.

In the region above the node where new tissues are being formed in the growing internodes the xylem consists entirely of ringed and spiral vessels, there being no pitted vessels. The sclerenchyma of the older parts of the in-

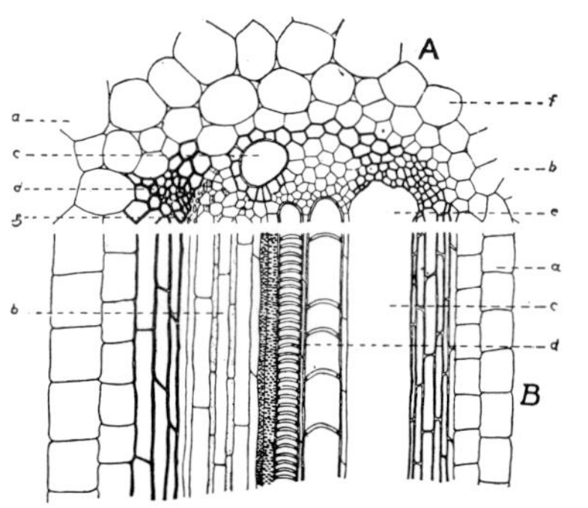

1

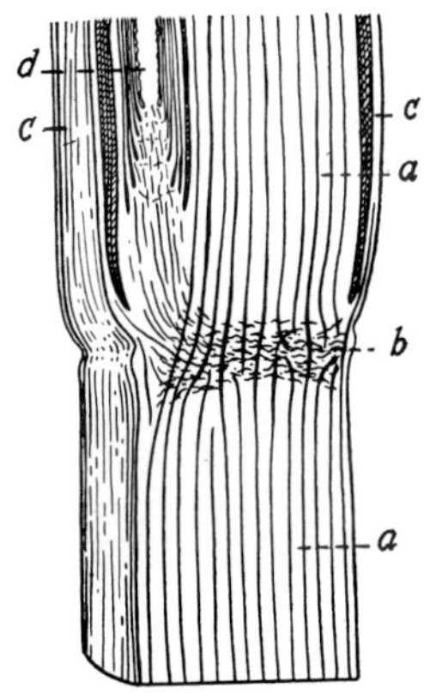

2

FIG. 17.—1, two-plane view of vascular bundle of stem. X 233. A, cross section. a, side of bundle toward outside of stem; b, side of bundle toward inside of stem; c, large pitted vessel; d, supportive tissue; e, lacuna; f, parenchyma cells of pith; g, compressed protophloem. B, longitudinal section. a, parenchyma cells of pith; b, phloem cells of bundle; c, lacuna; d, xylem elements of bundle.

2, vertical section of portion of a corn stalk, including an ear node, showing arrangement of vascular bundles. a, internodes; b, node; c, leaf sheath; d, ear shoot. (Diagrammatic.) X 1.

ternode is here represented by collenchyma. The corn stem is much like that of sugar cane described by Artschwager *(4)*.

If a fairly well developed piece of stalk containing a node is placed for a prolonged period in water and allowed to partially decay, i.e., is retted, the parenchyma tissue is destroyed while the bundles are left intact and can readily be separated. This is a satisfactory method for studying the course of bundles through the node and the cross connections and branching of the bundles. It is not satisfactory in very young stages in which the bundles are not yet sufficiently lignified to resist decay during such treatment. In internodes which are still growing rapidly the bundles in the region of new tissue just above the node will also decay and break because of their meristematic condition, so that it is not possible to follow bundles through several nodes. For tracing individual bundles through the stalk the method of staining before retting as suggested by Evans *(26)* may be more satisfactory. The cross connections in the node are differentiated later than the vertical bundles so that they may consist of a cylindrical mass of meristematic tissue after the vertical bundles have well differentiated tissues. At this stage they would also decay in the retting process.

The Leaf

GENERAL MORPHOLOGY

Leaves of the embryo.—Each true leaf of the embryo starts as a ridge around the developing stem just below its hemispherical growing point. This ridge grows upward and soon extends beyond the stem tip. As the leaf increases in width its edges overlap and it becomes rolled up into a hollow cone (Fig. 18), enclosing the stem tip and whatever higher, younger leaves may have formed, and is itself enclosed in any lower leaves and in the coleoptile.

The leaf becomes broader above than at its base where attached, so the veins spread out like a fan (Fig. 19–1) instead of being parallel as they will be in the fully formed leaf. When the stem increases in diameter the basal part of the leaf also grows in width, causing the veins to become parallel.

About five leaf initials are formed in the embryo before it becomes dormant at maturation of the kernel. This number does not include the first two modified leaves—the scutellum and coleoptile.

Leaves of the seedling and mature plant.—Upon germination the leaves already formed in the embryo resume growth, and the formation of leaf initials is resumed and continues until all the leaves have started and the growing point of the stem begins to differentiate to form the tassel. The total number of leaves formed varies both within and between varieties; the stalks of some locally adapted sorts were found to average 17 leaves attached above ground and 6 under ground. Not all of these are present as functioning leaves at any one time as some of the lower are lost before the upper ones have ex-

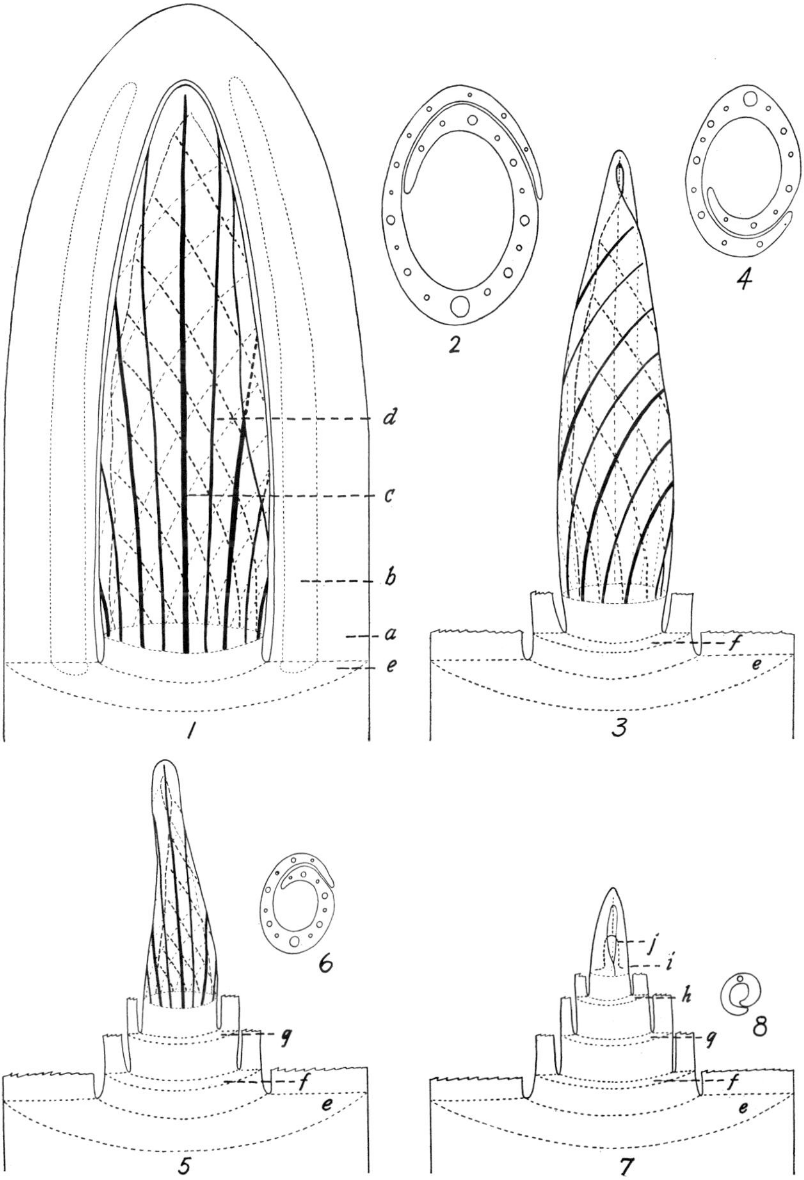

FIG. 18.—Dissections of plumule to show attachment and arrangement of the true foliage leaves of the embryo (diagrammatic). X 40.

1, first leaf of plumule exposed by partial removal of coleoptile; 3, 5, and 7, second, third, and fourth leaves exposed, respectively; 2, 4, 6, and 8, cross sections of first, second, third, and fourth leaves.

a, coleoptile; b, vascular bundle of coleoptile; c, midrib of first leaf; d, veins; e, coleoptile attachment; f, first leaf attachment; g, second leaf attachment; h, third leaf attachment; i, fourth leaf; j, growing point of stem.

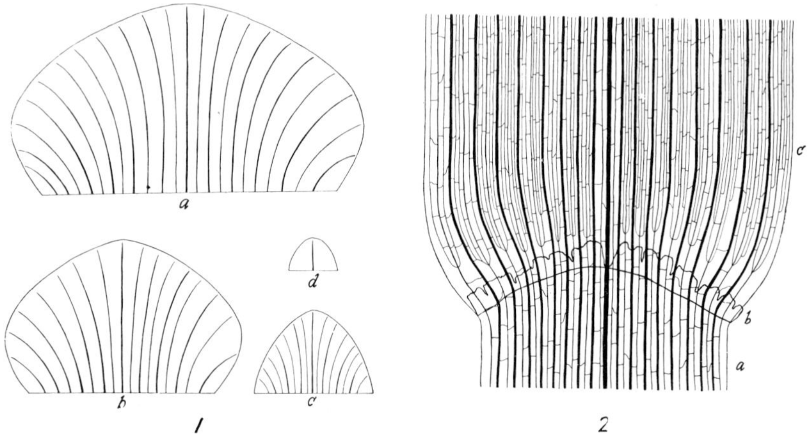

FIG. 19.—Comparative development of leaves in the embryo and the growing plant.
1, leaves of the plumule removed and unrolled, showing veins and relative size. X 12. a, first leaf; b, second leaf; c, third leaf; d, fourth leaf.
2, base of growing leaf, showing region where sheath and blade join, sometimes called the blade joint. a, upper part of sheath; b, ligule; c, blade, showing midrib and veins with branches and cross connections. X 4.

panded. The lower leaves tend to be torn loose and destroyed by formation of the crown roots and enlargement of the stem. Under favorable conditions the blades of the surviving leaves of full-grown stalks total about 1,400 square inches in area, or double this if both surfaces are considered.

One leaf is formed at each node. The successive leaves alternate on opposite sides of the stalk. Each leaf is rolled around the stem in the opposite direction from that of the preceding one. If in one the right-hand side overlaps, in the next the left-hand side will overlap.

The attachment of the leaf is not a circle but a short spiral extending a little more than once around the stem. The edge of the leaf which is attached lower on the stem overlaps the one attached higher. If one considers a plane passed vertically through the stem and leaves, so as to pass through all the midribs of the leaves, then all the halves which are attached lower and overlap the other half will be on the same side of the plane.

Parts of a leaf.—Each leaf consists of a thin, flat, expanded blade with a definite midrib and smaller veins, and a thicker, more rigid sheath with a less conspicuous midrib. Each sheath surrounds the internode above the node to which it is attached. Before the internodes elongate the lower sheath encloses the sheaths of leaves from higher nodes. The young, growing internodes alone would be too weak to hold the plant upright, and the strength of the young plant depends mainly on the stiff leaf sheaths. On the inner surface of the leaf where the blade and sheath join is a small collar, the ligule (Fig. 19–2), which extends upward surrounding the stem. The ligule in corn consists of a thin hyaline membrane without vascular tissue or green color. It is frequently said to prevent the entrance of rain into the sheath, but perhaps it would be as well to say it reduces the loss of water by evaporation from the space between the internode and sheath.

Just above the attachment of the ligule is a region formed by secondary intercalary growth. At first it appears as a light streak extending across the upper surface of the midrib to the edges of the leaf. Growth is more rapid at the edge of the blade than at the midrib, so that finally a V-shaped region known as the auricle is formed on each half of the blade, with its point toward the midrib. This region is sometimes called the blade joint. This joint causes the blade to bend outward and provides for its free lateral movement without tearing. The triangles of new tissue appear white or light green at first, but become discolored, brownish, and of a leathery texture in the mature leaf.

Development of the sheath and blade of the leaf.—The part of the leaf which is formed first is the tip of the blade. The growing region where new cells are formed remains at the base of the leaf. After the blade is formed the ligule is differentiated near the leaf base, thus making it possible to tell where blade and sheath meet. The formation of new cells at the base of the leaf continues, now forming the tissues of the sheath.

Later increase in size of the leaf blade is largely due to a rapid increase in size of the small meristematic cells which had previously been formed at its base. There is some cell division above the base, associated with the formation of vascular tissues and stomata. After the blade is nearly full sized it is still partially rolled up and full display is brought about partly by elongation of the internode beneath it and partly by elongation of the sheath.

The fully formed leaf consists mainly of the blade which is thin except for the midrib, and the thicker sheath whose midrib gradually disappears toward the base. The blade tapers gradually toward the tip and slightly toward the base until it abruptly narrows where it joins the sheath. Both blade and sheath have parallel veins which are united by cross connections at irregular intervals.

STRUCTURE

Arrangement of tissue.—Sections of the leaf (Figs. 20 and 21) show that it consists of an upper and lower epidermis between which is the mesophyll consisting largely of parenchyma cells with chloroplasts.

Embedded in the mesophyll are the veins of vascular bundles. There are two distinct types of bundles in the leaf, the larger primary bundles alternating with several smaller secondary bundles. Often about every tenth bundle is large, those between being much smaller. The primary bundles (Fig. 21–1 and 3) are similar to the bundles of the pith in the stem which have been described previously, and extend through about three-fourths of the thickness of the leaf. The remaining region between bundle and epidermis

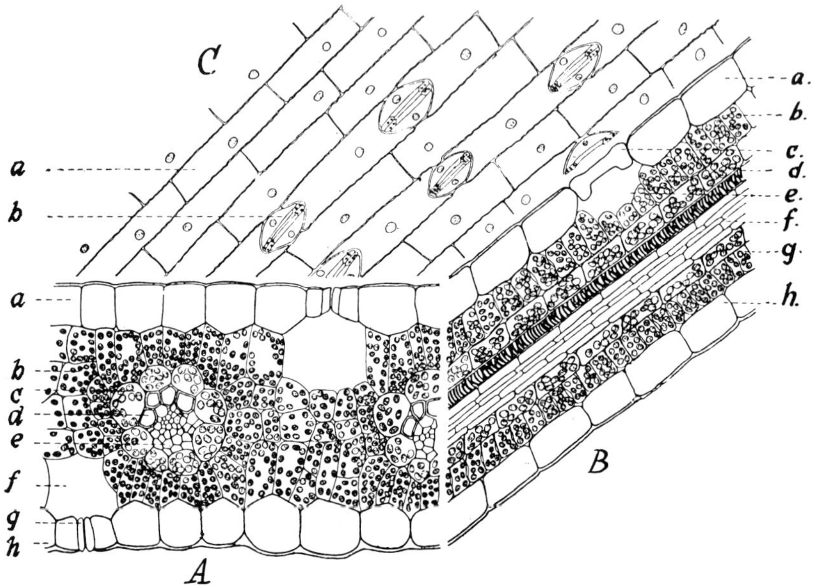

FIG. 20.—Cross and longitudinal sections and surface view of corn leaf, assembled to illustrate structure in three planes (diagrammatic and not drawn in perspective). X 233.

A, cross section. a, upper epidermis; b, chlorophyll-bearing tissue; c, chlorophyll-bearing bundle sheath; d, xylem of vascular strand; e, phloem of vascular strand; f, air chamber beneath stoma; g, cross section of stoma; h, lower epidermis.

B, longitudinal section. a, upper epidermis; b, chlorophyll-bearing tissue; c, longitudinal section of stoma; d, xylem of vascular strand; e, air chamber beneath stoma; f, phloem of vascular strand; g, chlorophyll-bearing bundle sheath; h, lower epidermis.

C, surface view of upper epidermis. a, epidermal cell with nucleus; b, stoma.

is occupied by sclerenchyma tissue. There is thus a strip without green tissue in the mesophyll at each of the larger bundles.

The smaller bundles are not accompanied by such fibrous tissue but are surrounded completely by a layer of sheath cells containing numerous large chloroplasts. Both tracheary vessels and sieve tubes with their companion cells are smaller and less numerous in the smaller veins than in the larger veins.

The epidermis.—The epidermis, which is but a single layer of cells in thickness, consists mostly of cells elongated (Fig. 21–4 and 5) parallel to the veins. At intervals on the upper surface of the leaf are bands of bulliform cells (Fig. 21–1), each band consisting of several rows of epidermal cells. Some of the adjacent epidermal cells develop hairs. By swelling, the bulliform cells cause the young leaf to unroll, and later when they lose their turgor cause the leaf to roll again. In rolling some of the stomata are covered, which is said to reduce transpiration. Since stomata occur on both leaf surfaces, they are by no means all covered by rolling. In some xerophytic grasses which have all their stomata in grooves on the upper surface such rolling may be a very effective means of reducing transpiration.

Distribution of stomata.—Through the stomatal aperture the air in the intercellular spaces of the mesophyll is connected with the outer air. By diffusion of gases through the stomata, carbon dioxide enters and is used

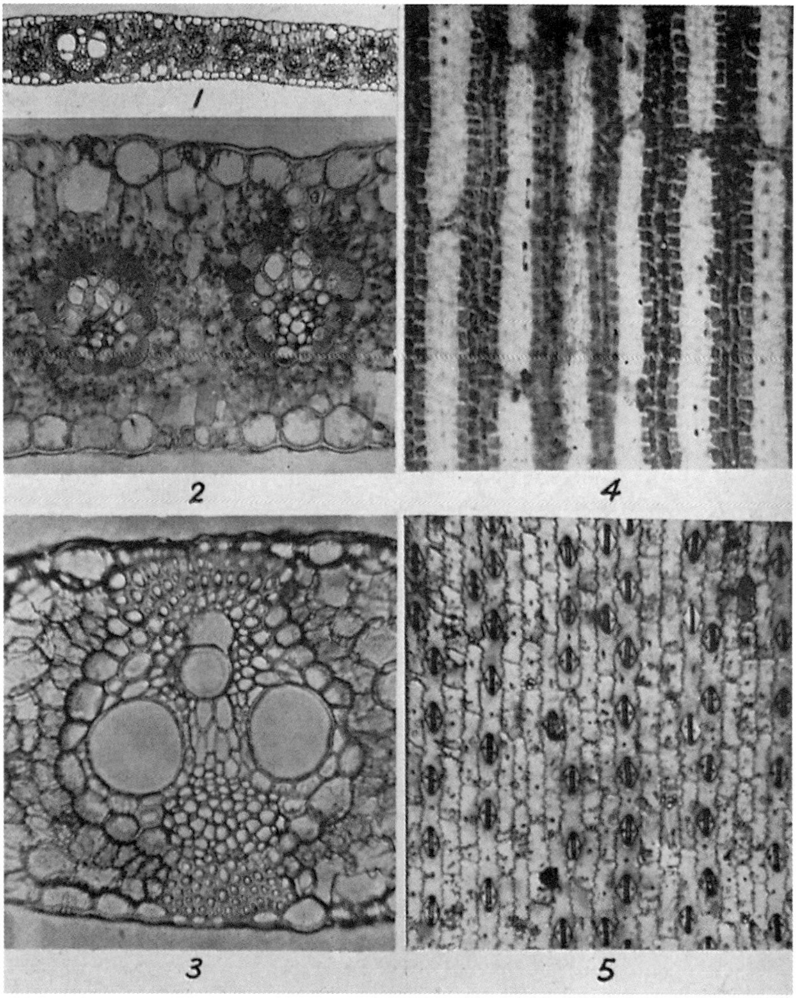

Fig. 21.—Photomicrographs of corn leaf.
1, cross section showing both large and small vascular strands. X 60.
2, cross section showing small vascular strands. X 250.
3, cross section showing large vascular strand. X 320.
4, veins with their cross connections (transparency through epidermis). X 100.
5, epidermis showing distribution of stomata. X 100.

in photosynthesis in the mesophyll while the oxygen set free in this process and carbon dioxide released in respiration escape along with water vapor. Stomata are found on both surfaces of the leaf blade, but are more numerous on the lower surface. On the sheath they are limited to the outer surface and they do not occur over the primary veins of the leaf where green parenchyma is lacking.

The number of stomata per unit area of the epidermis of the corn leaf varies considerably, but a large number of determinations *(38)* have averaged approximately 50,000 per square inch on the upper leaf surface and and 60,000 per square inch on the lower surface.

Origin and development of stomata.—At the time the stomata are about to be formed the epidermal cells have elongated until they are several times as long as wide. In each epidermal cell which is to form a stoma most of the cytoplasm and the nucleus move to the upper end of the cell. The nucleus divides and a cross wall is formed which divides the original cell into an upper short cell, only about as long as wide, and a lower elongated cell which is several times as long as wide, and which develops as an ordinary epidermal cell. The shorter upper cell divides longitudinally to form the two guard cells of the stoma. As the stoma grows less in width than the adjacent epidermal cells in the same row, the epidermal cells on each side bulge out to fill this space, thus forming a projection. The nuclei move into these projections and divide, walls are formed cutting off the projections as small cells which become the subsidiary cells of the stomatal apparatus, while the larger cells formed develop as ordinary epidermal cells.

The guard cells become thick walled and rigid except at the ends, where they remain thin walled and contain chloroplasts. The double walls between the guard cells separate into two separate walls, except at the ends of the cells *(15* and *83)*. When the guard cells become turgid their ends swell, thereby opening the aperture between their rigid central parts. When they lose their turgor they shrink and the stomatal aperture closes.

The stomatal aperture leads to an air cavity which extends under both guard and subsidiary cells. The size of the stomatal apparatus (Figs. 20 and 21) averaged about 50 x 38 microns in five eastern Nebraska types of corn grown at Lincoln *(44)*. The stomatal aperture averaged 17.1 microns long. Calculated as 3 microns wide when open, the area of the aperture would be 51.3 square microns. In these corn types there were an average of 78.7 stomata per square millimeter in the upper epidermis and 94.6 in the lower. The area of the apertures of all the stomata would thus be 0.404 per cent of the area of the upper epidermis, and 0.485 per cent of the area of the lower epidermis.

The average size of the stomata was less and the number per unit leaf area was greater in corn grown under the less favorable conditions of western Nebraska, whereas they were somewhat larger and the number per unit area somewhat less in corn grown farther east where growth conditions are still more favorable. There are also seasonal and varietal differences in size of stomata and in number per unit area.

DEVELOPMENT AND STRUCTURE OF THE REPRODUCTIVE ORGANS

Corn being monoecious, with staminate flowers in the tassels and pistillate flowers on the ear shoots (Fig. 1), it seems desirable to follow the development of each kind of inflorescence separately. The character of their growth is indeterminate. The main stem terminates in a staminate inflorescence or tassel, as do the basal branches or tillers when present. The tillers may be very much like the main stalk but their tassels frequently produce some pistillate flowers, mostly in the basal part of the tassel.

The branches arising from nodes above the soil surface terminate in a pistillate inflorescence or ear, but usually all soon degenerate except the upper one or two located about midway on the stalk. The amount of growth before disintegration varies as the buds may either die without much develop-

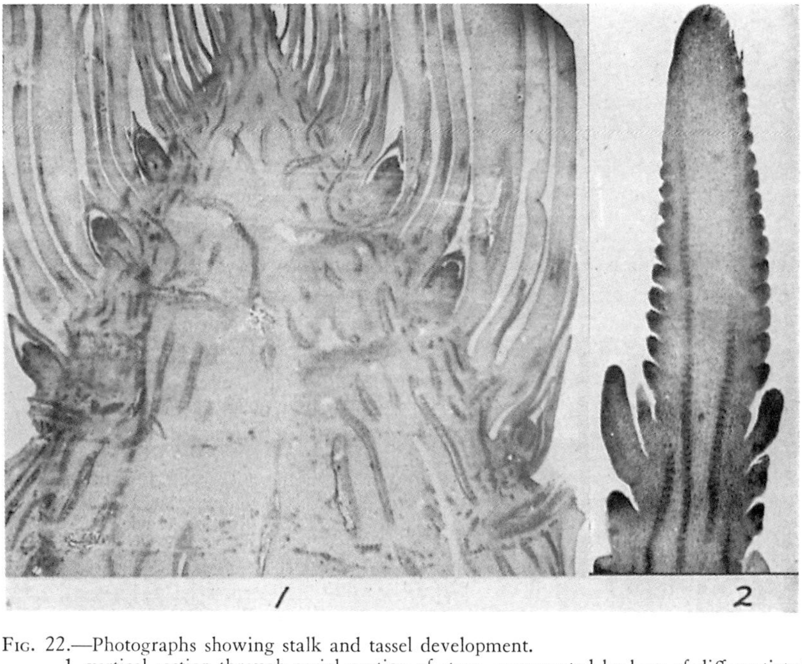

Fig. 22.—Photographs showing stalk and tassel development.

1, vertical section through aerial portion of stem, surmounted by base of differentiated tassel, 23 days after planting. Ear branches shown in leaf axils up to the uppermost ear shoot. X 25.

2, vertical section through tassel at stage of "1," showing basal branch initials, and spikelet initials on central spike. X 66.

FIG. 23.—Upper portion of stem, including uppermost ear branch, surmounted by tassel, 28 days after planting; nine internodes with leaf sheaths removed showing above the uppermost ear shoot. Stem height 3 inches above crown, tassel length 3/16 inch, plant height 21 inches. X 5.

ment or form ears several inches long which resemble the functional ears until they die. The morphological development of the inflorescences has been described by Bonnett (*8* and *9*).

Development of Tassel and Staminate Spikelets

The first indication of the differentiation of the tassel is that the growing point of the stem, which has been approximately hemispherical, begins to elongate and the initials of the tassel branches soon appear near its base as shown in Figure 10. This stage may, under favorable conditions, be reached in about two weeks after seedling emergence. The stem from the base of the crown to the tassel initial is only about an inch long at this stage, and the tassel about a millimeter long. Thus, three weeks after planting, the entire stem surmounted by the differentiated tassel may have formed under ground. The further growth of the tassel and stalk is illustrated in Figures 22, 23, and 24 and a mature spikelet is shown in Figures 25 and 27.

At first both the central axis and branches of the tassel are smooth, but outgrowths soon appear which become two lobed, each lobe finally giving rise to a spikelet with two flowers. The two glumes are formed in the same manner as ordinary leaves. In the axil of the lower glume a growing point forms, from which the lower flower is developed, while the original growing point of the spikelet gives rise to the upper or terminal flower. Each flower initial soon gives rise to a lemma or flowering scale, on the side adjacent to the glume. Later a palet forms opposite the lemma of each flower, the two palets thus being adjacent to each other between the two flower initials. Each growing point now gives rise to three stamens and two lodicules. One stamen in each flower is attached over the middle of the lemma, with a lodicule on either side of it, while the other two stamens are attached near the opposite edges of the palet. Each growing point then differentiates to form a rudimentary pistil (Fig. 26), which however does not develop far before its growth is arrested and it degenerates.

FIG. 24.—Photograph. Tassels at various stages of development. X 0.2.
1, six weeks after planting.
2, seven weeks after planting and in the reduction-division stage.
3, eight weeks after planting.
4, nine weeks after planting and in the pollen-shedding stage.

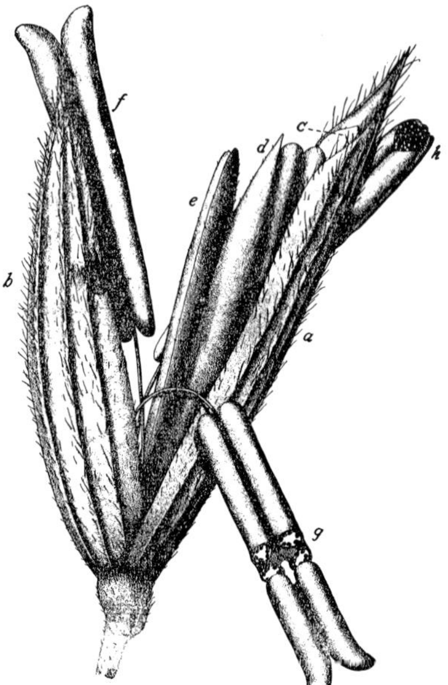

FIG. 25.—Complete staminate spikelet at pollen-shedding stage. a, lower or outer glume; b, upper or inner glume; c, lemma of lower flower; d, palet of lower or branch flower; e, palet of terminal flower; f, anther; g, cross section of anther; h, mature pollen. X 15.

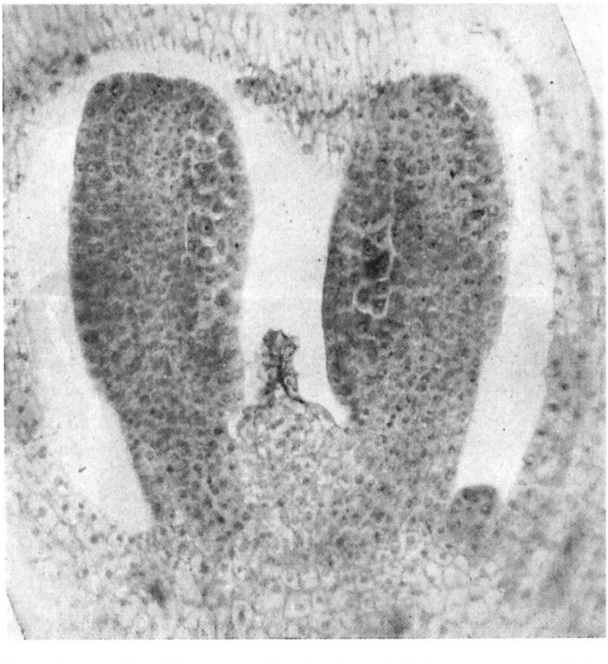

FIG. 26.—Photomicrograph. The normally abortive pistil between two of the developing stamens of a typically staminate flower at an early stage. X 150.

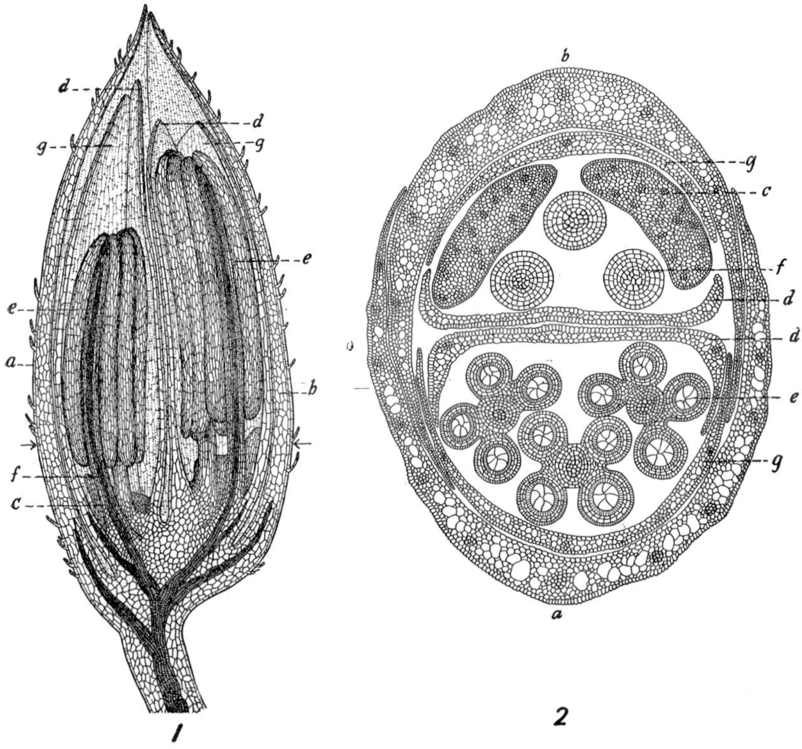

FIG. 27.—Organization and structure of staminate spikelet.

1, longitudinal, perspective view of both flowers, cut medially (one and one-half anthers of each flower have been removed). Right, upper or terminal flower; left, lower or branch flower. X 16.

2, cross section of spikelet cut at point marked by arrow in figure at left. Above, upper or terminal flower; below, lower or branch flower. X 32.

a, lower or outer glume; b, upper or inner glume; c, lodicules; d, palet; e, anther; f, filament of stamen; g, lemma.

THE STAMENS

The glumes, lemmas, and palets are protective organs for the stamens. The lodicules swell and thus pry the lemma and palet apart, enabling the stamens to be pushed out by the elongating filaments when the pollen is mature. The stamens are the only organs of the staminate flowers whose development it seems necessary to follow in detail. Their structure and development in corn is similar to that in other grasses which have been studied. The internal arrangement of the stamens and other organs within the staminate spikelet is shown in Figure 27.

Differentiation of tissues.—Each stamen begins as a lobe, the terminal part of which enlarges to form the anther, while the lower part forms the filament. In the development of each of the four chambers or loculi (Fig. 28) of an anther, there is a central row of archesporial cells which give rise to a column of sporogenous tissue and a three-layer wall between it and the enclosing epidermis. At each division in the formation of the anther wall, it is the inner layer of cells which divides. After the last division the innermost layer becomes the tapetum. The cells of this layer are filled with food that is later used by the developing spore mother cells.

The cells of the sporogenous tissue also divide until a large number of spore mother cells are formed. At first these fill the cavities of the anther but later, as the cavity enlarges, a vacuole forms in the center around which

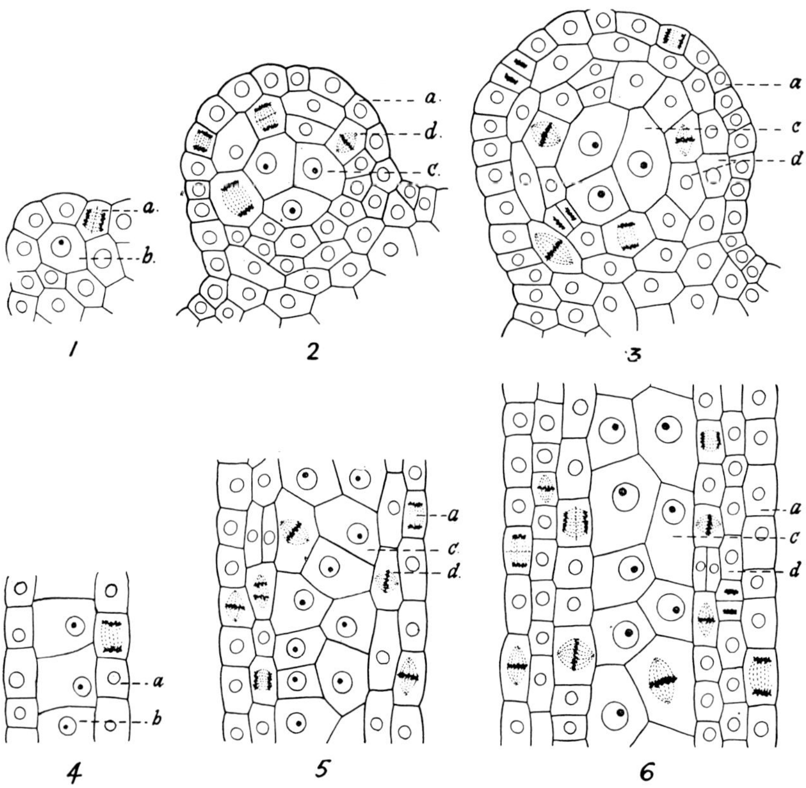

FIG. 28.—Stages in the development of one of the four loculi or chambers of an anther, showing diagrammatically the cell division during growth of the epidermis and of the sporangium which consists of the sporogenous tissue and the wall between it and the epidermis. a, epidermis enclosing sporangium; b, archesporium which is a single longitudinal column of cells, each of which divides laterally, one row of daughter cells giving rise to sporogenous tissue and the other to the wall layers; c, sporogenous tissue from which the microspore mother cells originate; d, wall of sporangium.

1, 2, and 3, cross sections of loculus at early, medium, and late stages. X 500.

4, 5, and 6, longitudinal sections at stages corresponding to 1, 2, and 3. X 500.

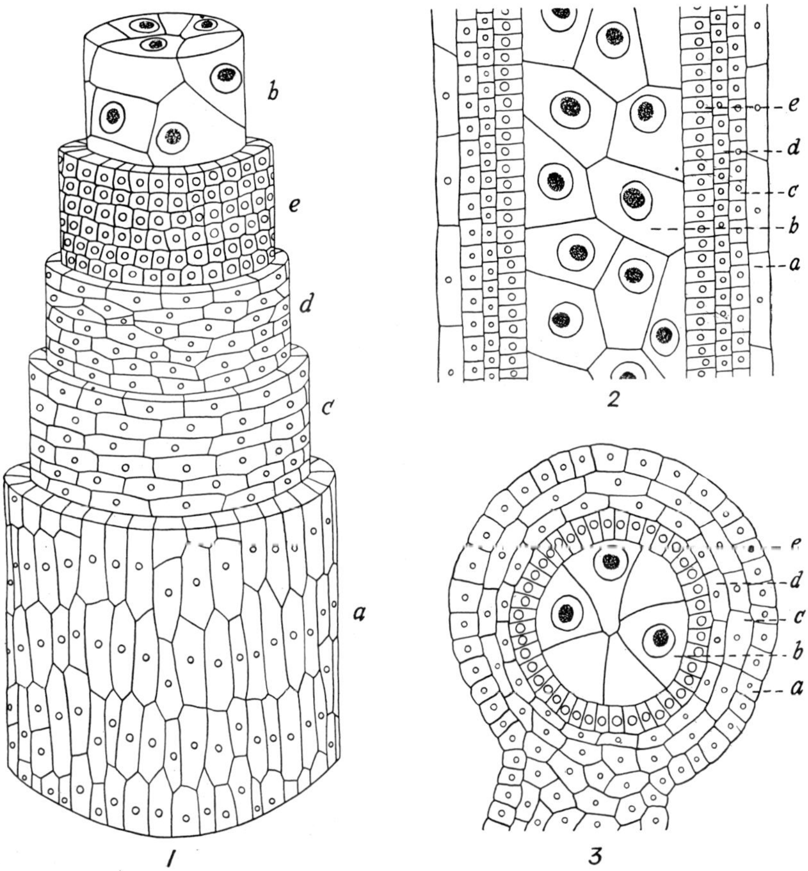

FIG. 29.—Structure of one loculus of anther, showing central column of sporogenous tissue and the three layers of the wall between it and the epidermis. a, epidermal layer; b, sporogenous tissue which gives rise to the microspore mother cells; c, d, and e, outer, middle, and inner (tapetal) layers of sporangium wall. X 220.

1, longitudinal, perspective view of loculus with successive portions of the epidermis and three wall layers removed.

2 and 3, longitudinal and cross sections of same, respectively.

the spore mother cells form a single layer of cells lining the cavity. The cell divisions involved in the growth of the anthers are illustrated in Figure 28, and structure of the anther wall enclosing the column of sporogenous tissue is shown in Figures 29 and 30.

Reduction-division.—The spore mother cells now undergo division (Figs. 31 and 33). Each of them contains two sets of 10 chromosomes, one set from each parent. During synapsis, which is an early stage of reduction-division or meiosis, each chromosome of one set pairs with the homologous or corresponding member of the other set. The 20 single or univalent chromo-

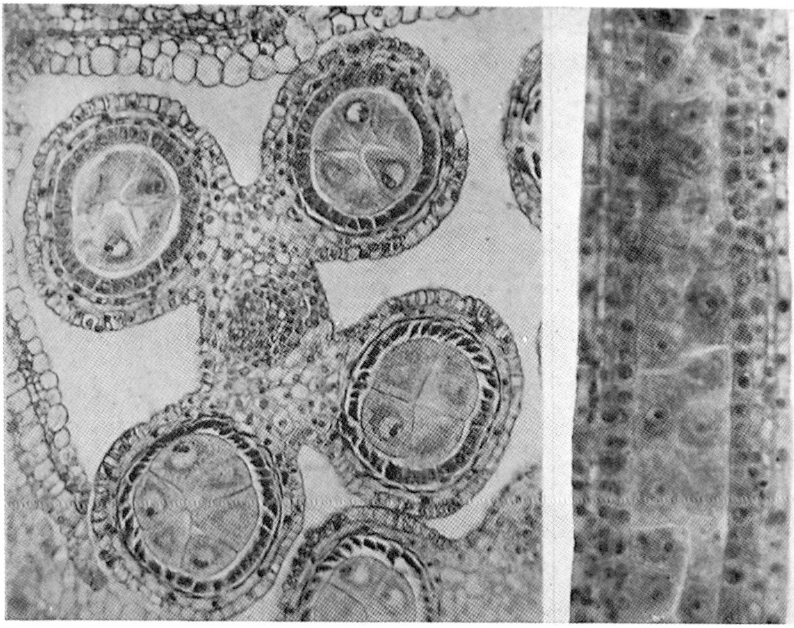

FIG. 30.—Photomicrographs. The anther, showing microspore mother cells before meiosis. X 125.

Left, cross section of anther; right, longitudinal section of part of one loculus of anther.

somes thus form 10 double or bivalent chromosomes. Each strand of these bivalent chromosomes splits to form two strands, so that there are four strands or chromatids in each bivalent chromosome. It is in the four-strand condition that *crossing-over* takes place, forming new chromatids by exchange of segments derived from each parent.

While in this early prophase stage the chromosomes are long, slender, and tangled, and may be most readily studied with respect to their detailed morphology (*58* and *59*). It has been shown that each member of the complement of 10 chromosomes can here be recognized by its architecture (*55* and *60*). They now shorten until they may be only about twice as long as broad, a stage usually known as *diakinesis*. By this time crossing-over is completed. This is a favorable stage for counting as the short, thick chromosomes are much more easily counted than the long, tangled ones of earlier stages.

Two nuclear divisions follow in rapid succession, forming four nuclei in each spore mother cell, each with but half the number of chromosomes that were present in the original mother cell. In corn there are usually 20 chromosomes in the spore mother cell before reduction and 10 in each nucleus after reduction (*27, 45, 48, 49, 54, 70,* and *72*). Each of the four new nuclei have one of the four strands of each bivalent chromosome that were present

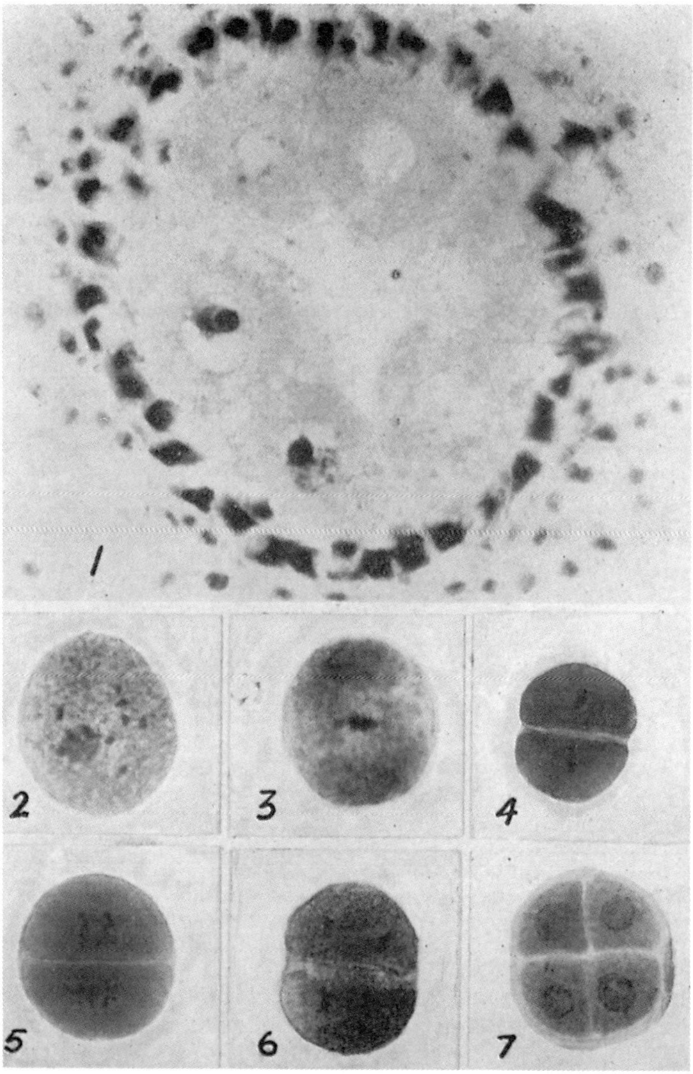

Fig. 31.—Photomicrographs showing microsporogenesis or reduction-division. X 350.

1-3, stages during the first meiotic division (1, leptonema or slender thread stage of chromosomes in microspore mother cells which are here surrounded by tapetal layer of anther; 2, diakinesis; 3, metaphase).

4-6, stages during second meiotic division (4, prophase; 5, metaphase; 6, telephase).

7, quartet or four-spore stage, each spore being an immature haploid, one-nucleate pollen grain.

at diakinesis. These processes do not occur simultaneously throughout the tassel, reduction-division beginning a little above the base of the central axis and extending both upward and downward, reaching the base long before it reaches the tip. Similarly reduction-division in the tassel branches begins a little above their bases and advances upward and downward.

Microspores or pollen grains.—A microspore is organized around each of these nuclei, each surrounded by a cell wall inside of the wall of the spore mother cell; this wall of the mother cell, however, soon disintegrates, leaving the microspores or pollen grains free in the cavities of the anthers. They are thin walled, and when set free from the mother cell assume a nearly spherical form. The walls now thicken but remain relatively thin at the single germ pore through which the pollen tube will emerge during germination of the pollen.

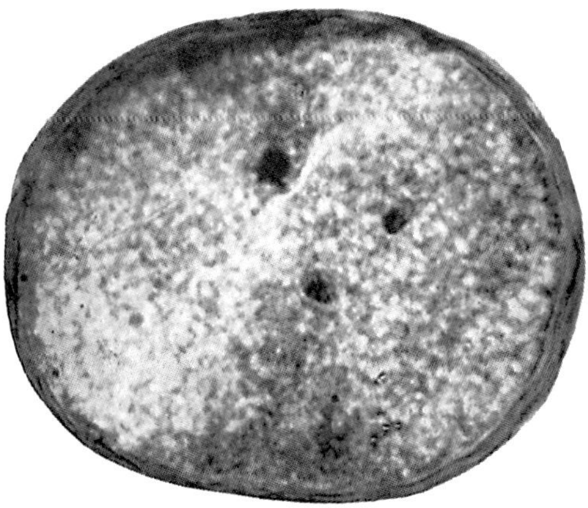

FIG. 32.—Photomicrograph. Pollen grain at three-nucleate stage just before its maturity. The three nuclei, each with 10 univalent chromosomes, have arisen by two divisions following the quartet stage shown in Figures 31—7 and 33—9. One of them remains as the vegetative nucleus and the other two become the crescent-shaped sperm cells as shown in Figure 33—10. X 750.

The pollen remains uninucleate until shortly before shedding. By this time the cytoplasm has become very dense, and is filled with a large number of very small starch grains, making it difficult to see the nuclei except in sectioned pollen. The nucleus then divides into a vegetative or tube nucleus and a generative nucleus which later divides to form two sperm cells (Figs. 32 and 33). The actual process of division was not seen but two and three nucleate stages were found.

Shedding of pollen.—At anthesis, just prior to shedding of pollen, the lodicules swell to several times their former size and pry the lemma and palet apart, making it possible for the anthers to be extruded by the elongating filaments. Soon afterward the anthers break open near the tip, forming pores

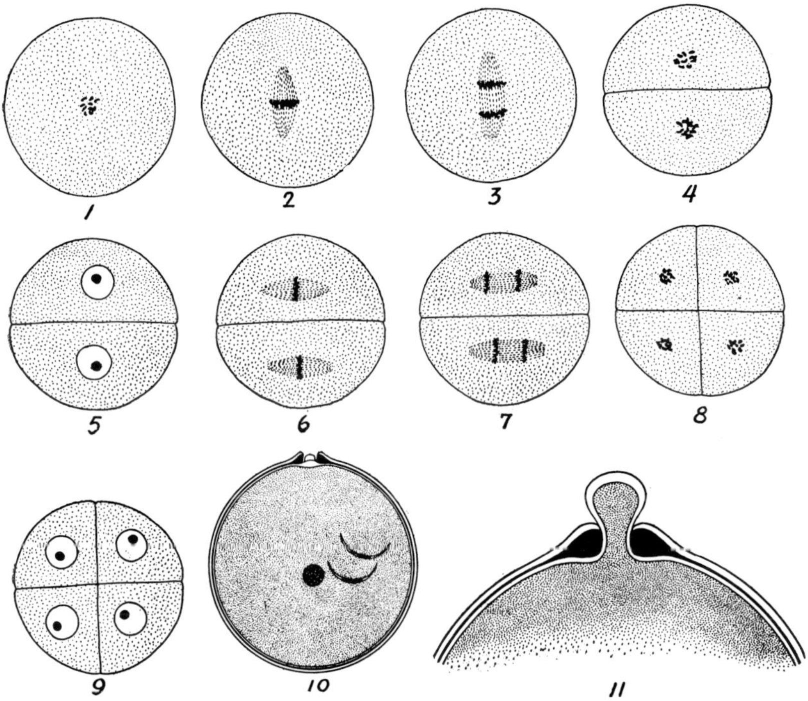

FIG. 33.—Development of pollen from the microspore mother cell by sporogenesis (diagrammatic), followed by internal differentiation of each of the one-nucleate spores into a three-celled male gametophyte.

1-5, stages during first meiotic division (1, diakinesis; 2, metaphase; 3, anaphase; 4, telephase; 5 two-cell stage). X 400 to 350.

6-9, stages during second meiotic division (6, metaphase; 7, anaphase; 8, telephase; 9, quartet or four-spore stage). X 350 to 300.

10, pollen grain showing exine and intine layers of wall, germ pore with plug, tube or vegetative nucleus, and two sperm cells. X 350.

11, portion of germinating pollen grain, showing formation of pollen tube by protrusion of intine. X 800.

through which the pollen escapes. Little pollen is shed until the anthers are shaken by the wind or otherwise disturbed, so that in general the pollen is retained until there is enough wind to carry it some distance from the plant by which it was produced. This tends to insure a high percentage of cross pollination under ordinary field conditions. In each plant the tassel usually sheds some of its pollen before the silks of its ears emerge from the husks, but the tassel normally continues to shed for several days after the silks are ready to be pollinated *(76* and *84)*.

A tassel begins to shed pollen a short distance below the tip of the central axis, and shedding progresses both upward and downward, reaching the tip of the central axis long before it reaches the base. The branches start shedding a little later than the central axis, the first spikelets to begin shedding

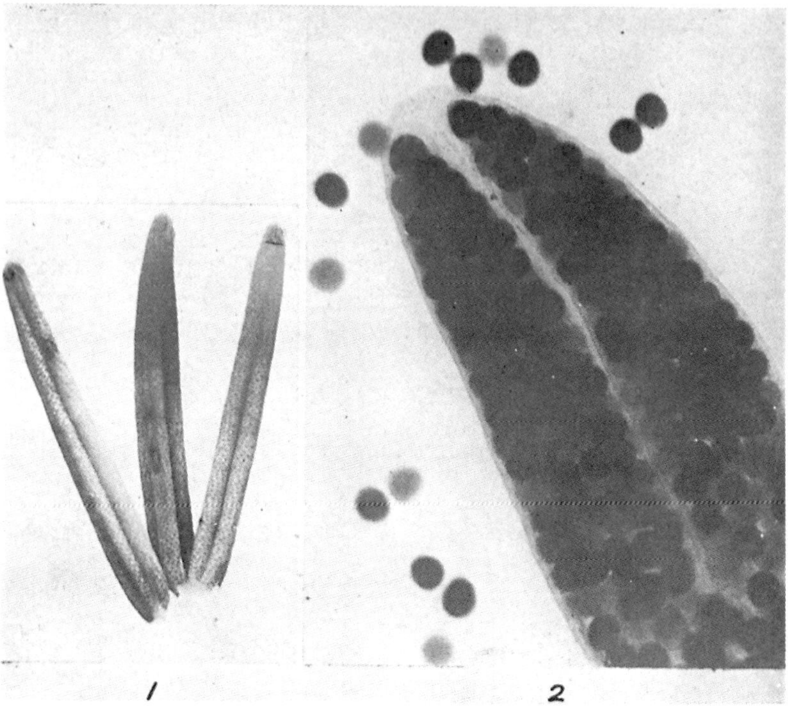

FIG. 34.—Photomicrograph. Mature anthers at pollen-shedding stage.
1, the three anthers of one staminate flower. X 10.
2, tip portion of anther at pollen-shedding stage, broken to show the mature pollen content. X 45.

being a little below the tip of each branch. The upper flower of each spikelet begins shedding pollen about one to three days before the lower. A tassel may shed pollen for a week or more *(89)*.

Amount of pollen produced.—Pollen grains are produced in enormous numbers (Fig. 34), as is the rule in wind-pollinated plants. Sturtevant *(84)* estimated that a tassel would produce about 18,000,000 pollen grains. An average-sized tassel of Nebraska White Prize corn grown at the Nebraska Experiment Station was calculated to produce about 25,000,000 pollen grains. On this basis 25,000 pollen grains are produced for each kernel on an ordinary ear with 1,000 kernels.

With a stand of three stalks in hills 42 inches apart, an area of 588 square inches is available in the field for each stalk. Thus an average of 42,500 pollen grains are provided for each square inch of the field. If the silks of an ear display a total surface of 4 square inches they will intercept about 170,000 pollen grains. Estimating 1,000 silks per ear, this amounts to 170 pollen grains per silk. Considering that a corn field sheds pollen for 13 days

(39), each silk receives an average of 13 pollen grains per day. Under favor-ble weather conditions, most of the silks are pollinated the first day after they

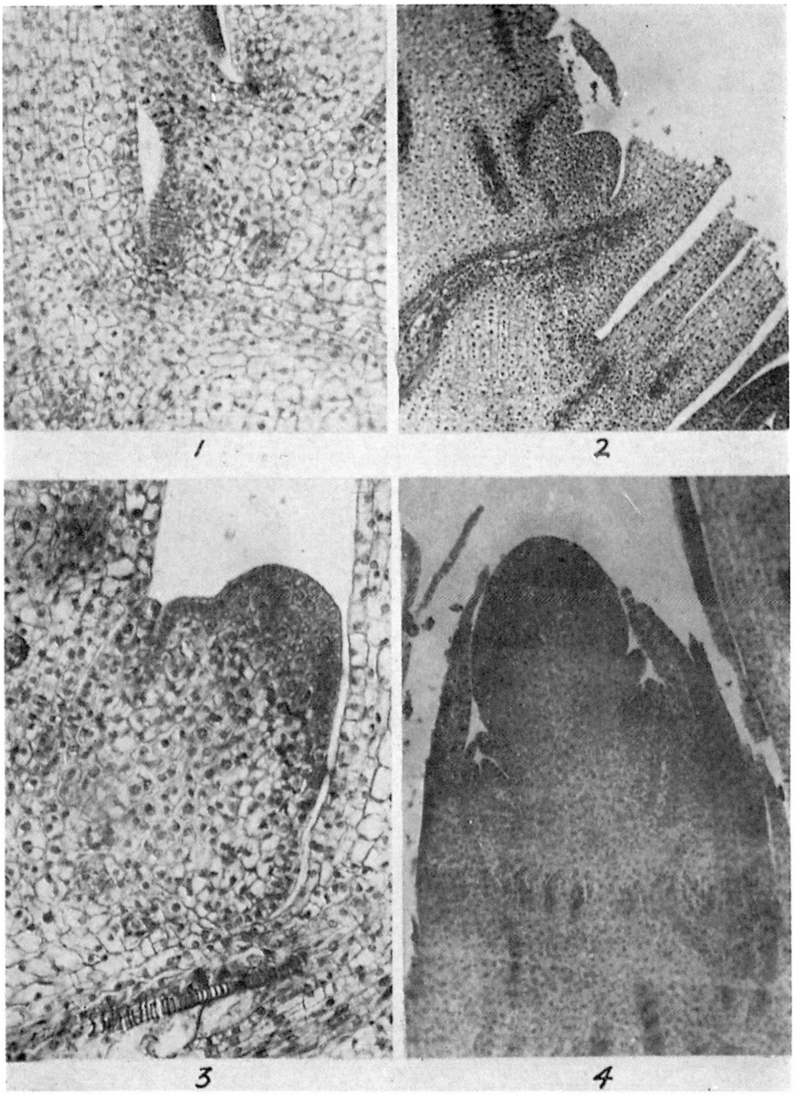

FIG. 35.—Photomicrographs. Early stages in the development of the ear shoot.

1, first trace of cell differentiation in formation of ear shoot in the axil of the leaf (left center). X 40.

2, the growing point takes on definite form. X 20.

3, several bulges or prophyllum and husk initials have started (a spiral vessel leading to the leaf shows prominently). X 60.

4, the young ear shoot with differentiated husks and smooth ear initial. X 32.

emerge, and the pollination of individual ears is largely completed over a three-day period.

FIG. 36.—Photograph. Ear shoots picked from nine successive nodes on a typical eastern Nebraska corn stalk growing under favorable conditions. More than the uppermost one or two ear shoots seldom emerge from beneath their leaf sheaths. Disintegration of the others begins at an early date.

Above, unhusked ear shoots; left to right, lowest to uppermost. X 0.15.

Below, ears from ear shoots shown above; left to right, lowest to uppermost. X 0.4.

Development of the Ear Shoot and Pistillate Inflorescence

As previously stated, an axillary branch bud forms at each node of the stem up to the one which bears the uppermost ear (Fig. 22). Each bud is enclosed in a leaf-like structure, the prophyllum. This differs from an ordinary leaf in having two main ribs instead of one midrib, and it usually lacks a ligule and blade. At first those buds which will develop into ear shoots (Fig. 35) are like those which form tillers, but differences soon appear. In tillers the prophyllum remains small and soon dies, whereas in ear shoots it becomes large and persists as one of the husks, often becoming about 9 inches long and 1½ to 2 inches wide. As shown in Figure 36, in cornbelt varieties only the upper one or two ear shoots reach the fertilization stage, although others may attain considerable size before disintegration.

FIG. 37.—Photograph. Pistillate branch or shank, terminated by ear in "roasting-ear stage." Husks partly removed to expose nodes, and husk and ear attachments, 23 days after pollination. X 0.2.

The branch or shank (Fig. 37) differs from ordinary stems in being more slender, the internodes generally remaining short, thereby enabling the leaf sheaths to surround the ear as husks, while the leaf blades remain small in most sorts of dent corn. The sheaths are much broader and thinner than the sheaths of ordinary leaves. The outer husks are distichous like ordinary leaves while the inner are polystichous, there sometimes seeming to be as many ranks as there are double rows of kernels. When all the leaf or husk initials have formed, the growing point of the ear shoot elongates to form the beginning of the ear. This is somewhat later than tassel differentiation. The early stages of the ear are like the corresponding stages described earlier in the development of the tassel except that there are no branches on normal ears.

At first the ear is smooth (Fig. 35–4) but protuberances soon form in rows (Fig. 38). The basal protuberances are formed first and development advances toward the tip of the ears. Each one becomes two lobed, each lobe developing into a spikelet with two flowers, only one of which commonly persists. Since each spikelet normally produces one kernel of corn the kernels also will be in double rows and there will be an even number of rows of kernels on the ear. The paired character of spikelets and kernels, and the organization

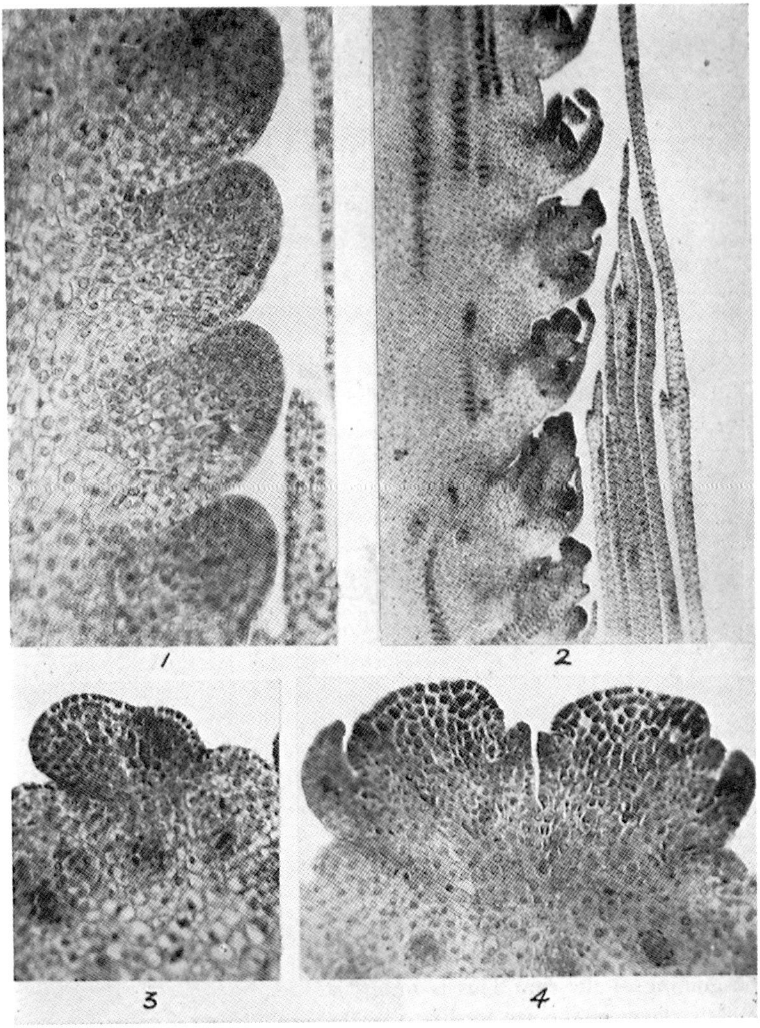

FIG. 38.—Photomicrographs. Early stages in the development of the pistillate spikelets on ear.

1, longitudinal section of ear showing smooth protuberances arranged in rows, each of which is the initial of a pair of spikelets. X 160.

2, longitudinal section of ear enclosed in growing husks, after the protuberances shown in "1" have become more differentiated as spikelets. X 40.

3, cross section of portion of young ear, showing the protuberances of "1" becoming two lobed. Each lobe develops into a spikelet. X 80.

4, two typical spikelet initials developed from a single original protuberance. This accounts for the paired arrangement of the rows of grain on an ear, as shown in Figure 39. Initials of the glumes and lemmas are shown, and slight bulges of growing points in the differentiation of additional organs are apparent. X 120.

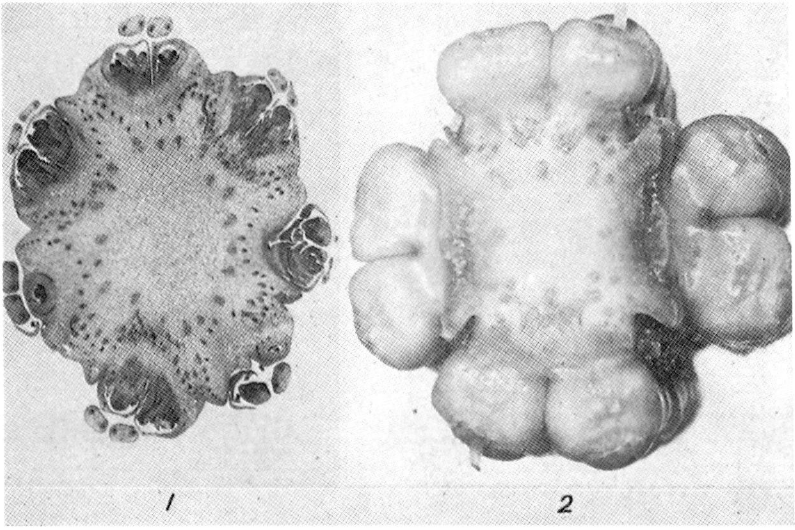

Fig. 39.—Photographs.

1, cross section of young ear showing paired arrangement of developing spikelets on the cob. X 7.

2, end view of a broken, young, fertilized ear of eight-row corn, showing the paired arrangement of rows of grain on the cob. Each kernel represents one original spikelet initial shown in Figure 38—4. X 2.

of tissues of the rachis or cob to which they are attached are shown in Figure 39. Alternation of pairs of spikelets in adjacent pairs of rows as related to cob structure is apparent in Figures 2 and 72. When each spikelet forms two kernels through functioning of both flowers, as in Country Gentleman sweet corn (Fig. 72, fifth ear from left in lower row), the rows become irregular due to crowding *(37, 41, 65, 82, 90* and *93)*. Descriptions of the morphology and anatomy of the corn cob are given by Lenz *(51)* and Laubengayer *(52)*.

THE PISTIL

Differentiation of the tissues.—The growing point of the upper flower, after giving rise to the initials of the other organs (Fig. 40), is differentiated to form the functional pistil whose further development is illustrated by photomicrographs in Figure 41. A ring-like outgrowth starts at its base which is generally considered to represent three undiverged carpels as found in many other monocotyledons.

The part above the attachment of the carpels develops a single sessile ovule which consists of a nucellus with two integuments or rudimentary seed coats (Fig. 41–1 and 2). There is no well defined funiculus or ovule stalk and any rudiment of such a structure is merged with the placental

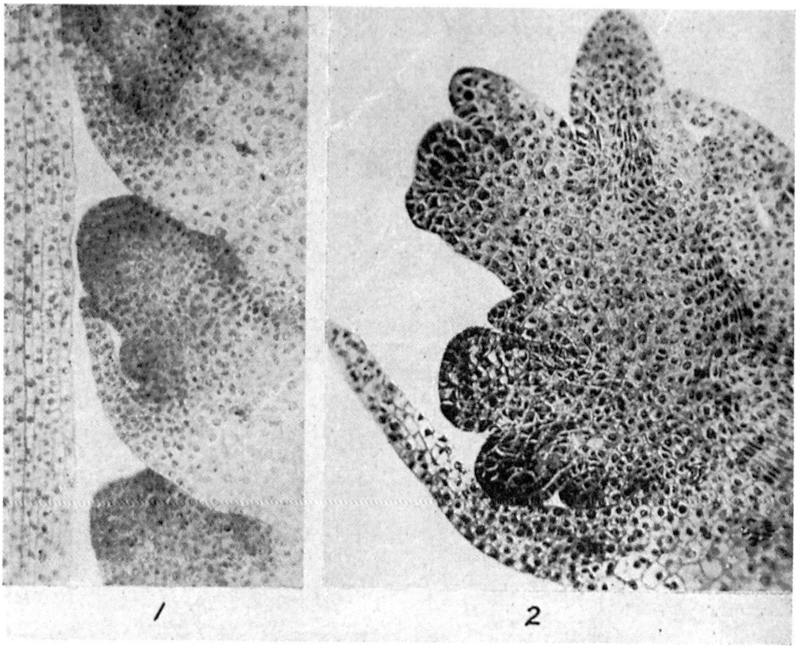

FIG. 40.—Photomicrographs. Young developing pistillate spikelets.

1, longitudinal section of spikelet at very early stage when differentiation of organs begins. X 160.

2, longitudinal section of a spikelet at stage when the initials of nearly all the important organs of the spikelet have been differentiated. From top to bottom: inner glume, lemma, stamen, one of the two anterior carpels which give rise to the silk and also join with the posterior carpel to form the ovary wall, growing point of upper flower which will terminate in the nucellus, posterior carpel, palet (slight bulge below posterior carpel), palet of lower flower, growing point of lower flower, stamen, lemma, and outer glume. X 160.

tissue at the broad, circular seat of ovule attachment. This region is co-extensive with the chalaza which is the location where the nucellus and integuments of the ovule are united.

The united carpels, which will form the ovary wall or pericarp of the mature kernel, grow upward until they completely enclose the ovule (Fig. 41–2). Where they meet, the functionless so-called stylar canal is formed *(31* and *86)*. The growth of both nucellus and integuments is more rapid on the posterior side of the ovule which thereby becomes approximately campylotropous. The inner integument grows until it forms a thin membrane covering the entire nucellus except for a small opening, the micropyle, just over the mature embryo sac (Fig. 41–4 and 5). The outer integument growing up from the posterior side does not extend as far, reaching only a little beyond the stylar canal into which it makes a small plug-like growth. Neither

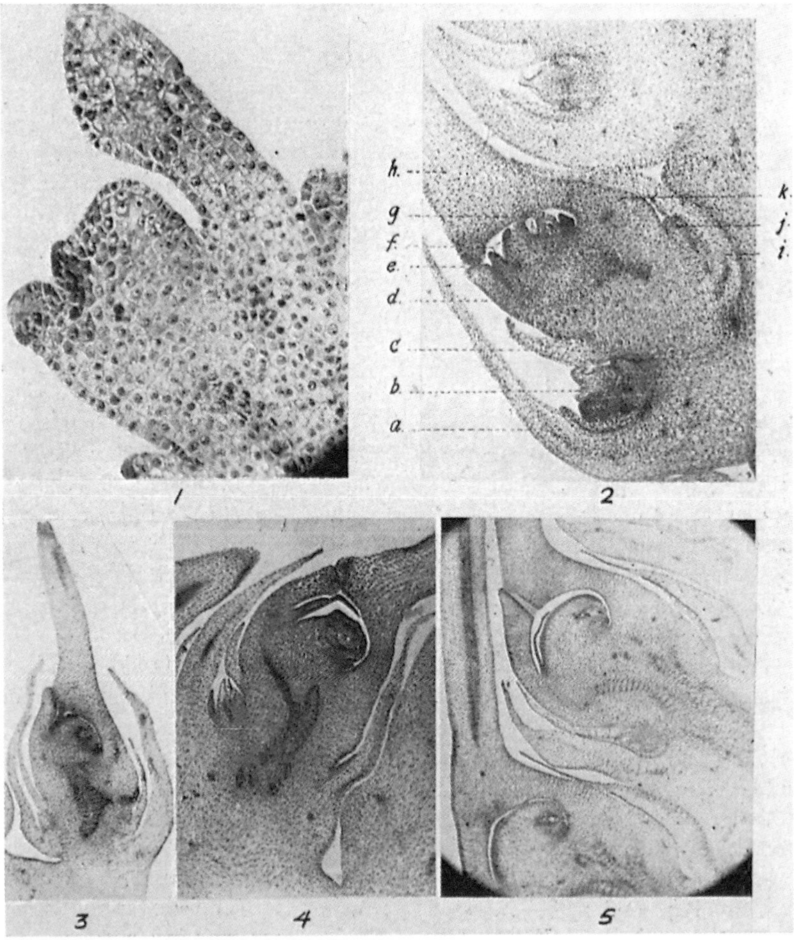

FIG. 41.—Photomicrographs. Medial longitudinal sections showing intermediate to late stages in development of pistillate spikelet, before and during sporogenesis and growth of the embryo sac. Note the more rapid growth of the seed coats or integuments and of the ovule on posterior side, carrying the embryo sac to a more basal position.

1, early stage of functioning flower previous to formation of megaspore mother cell showing, left to right: palet, posterior carpel, nucellus with seed-coat initials, one of anterior carpels, and lemma. X 140.

2, complete spikelet showing abortive flower and the functional flower with megaspore mother cell about to divide. a, outer glume; b, abortive flower with lemma and palet; c, palet of functional flower; d, posterior carpel; e, stylar canal; f, two seed-coat initials, showing also on anterior side of nucellus; g, nucellus with megaspore mother cell; h, style or silk; i, inner glume; j, lemma; k, anterior carpel. X 70.

3, four-spore stage following reduction-division. X 30.

4, two-nucleate embryo-sac stage. X 60.

5, nearly mature, eight-nucleate embryo-sac stage. X 40.

integument makes much growth below the micropyle, which becomes almost basal *(29, 65, 78,* and *88).*

While the carpels are growing, the two anterior ones which face the ear tip form outgrowths which develop into the style or silk (Fig. 41–2). A vascular strand passes up through each of these carpels into the silk, extending to the branched tip where one goes into each branch. Very rarely the posterior carpel also takes part in forming the silk (Fig. 49), which then has three strands of vascular tissue and three free tips instead of the usual two. Such three-parted silks are sometimes hollow, the stylar canal being continuous through the silk. The surface of the silk becomes covered with numerous hairs which are developed from cells of the epidermis. These are most numerous along the edges of the silk. The fully developed hair is composed of four rows of cells (Figs. 50 and 51), so arranged that the intercellular spaces between the rows form a continuous opening through the center of the hair. At the base of the silk is a growth zone where new cells develop, causing continuous elongation of the silk (Fig. 48–2) until it is pollinated and fertilization takes place. The zone of new growth then shrivels up, forming the place where the silk breaks loose. The silk is perhaps best considered as consisting of a very short style without hairs, above which is the region of intercalary growth. The branched part corresponds to the stigma of other grasses, and the part formed by intercalary growth is an addition to the stigma which has nothing corresponding to it in most grasses.

Reduction-division.—In the young nucellus, before the archesporium is differentiated, an axial row of cells (Fig. 42–1) is often noticed which terminates in a sub-epidermal cell somewhat larger than the surrounding cells. This terminal cell divides, forming cells at its base and sides until it enlarges and becomes the archesporial cell (Fig. 42–2). The surrounding cells continue to divide, while the archesporial cell enlarges and becomes the megaspore mother cell. The archesporium is a hypodermal cell overlaid by a single nucellar layer of epidermal cells until the beginning of meiosis or reduction-division (Fig. 42–3 to 8) corresponding with that described for the anther under *Differentiation of Tissues* on page 41. Through meiosis, a group of four megaspores end to end is formed, corresponding with the microspore quartet. Changes occurring in the ovule during meiosis and embryo-sac development [5] as shown in Figures 42 and 43 are illustrated in Figure 41.

The three megaspores nearest the micropyle soon degenerate and the innermost spore alone persists to function as the first cell of the embryo sac. These stages of megasporogenesis were missed by some of the earlier investigators, leading them to suppose that all the megaspores functioned in producing the embryo sac. During this interval of meiosis and embryo-sac

[5] In Figures 42 to 44, and 46 and 47, illustrating megasporogenesis and embryo-sac development, the outer or micropylar end is always shown facing in the same direction (downward) for the sake of clarity, although its position gradually becomes almost reversed during ovule development, as shown in Figure 41.

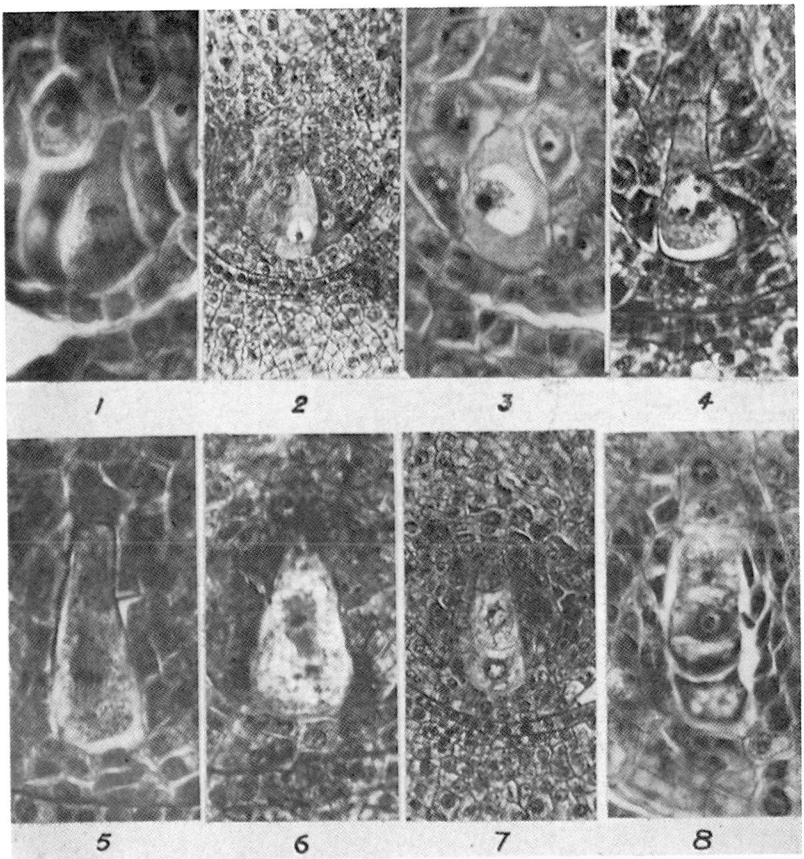

FIG. 42.—Photomicrographs. Longitudinal sections through part of nucellus before and during megasporogenesis.

1, hypodermal cell undergoing mitotic division before differentiating into the megaspore mother cell. X 500.

2, megaspore mother cell before meiosis begins. X 250.

3-7, stages during the first meiotic division (3, zygonema, X 500; 4, diakinesis, X 500; 5, metaphase, X 500; 6, telephase, X 500; 7, two-cell stage, X 250).

8, quartet of four-spore stage, formed by second meiotic division. The innermost spore normally functions as the haploid embryo sac, whereas the other three degenerate. X 500.

formation, the nucellar epidermis divides periclinally (Figs. 42–44) to form a layer five or six cells thick, thereby gradually embedding the embryo sac more deeply. According to Randolph *(68)*, this is the only place where periclinal division takes place in the epidermis of corn.

Development of the embryo sac.—The nucleus of the megaspore (Fig. 43–1) divides, and a vacuole forms between the two nuclei (Fig. 43–2) which come to lie at opposite ends of the cell. This cell now is the two-nucleate stage

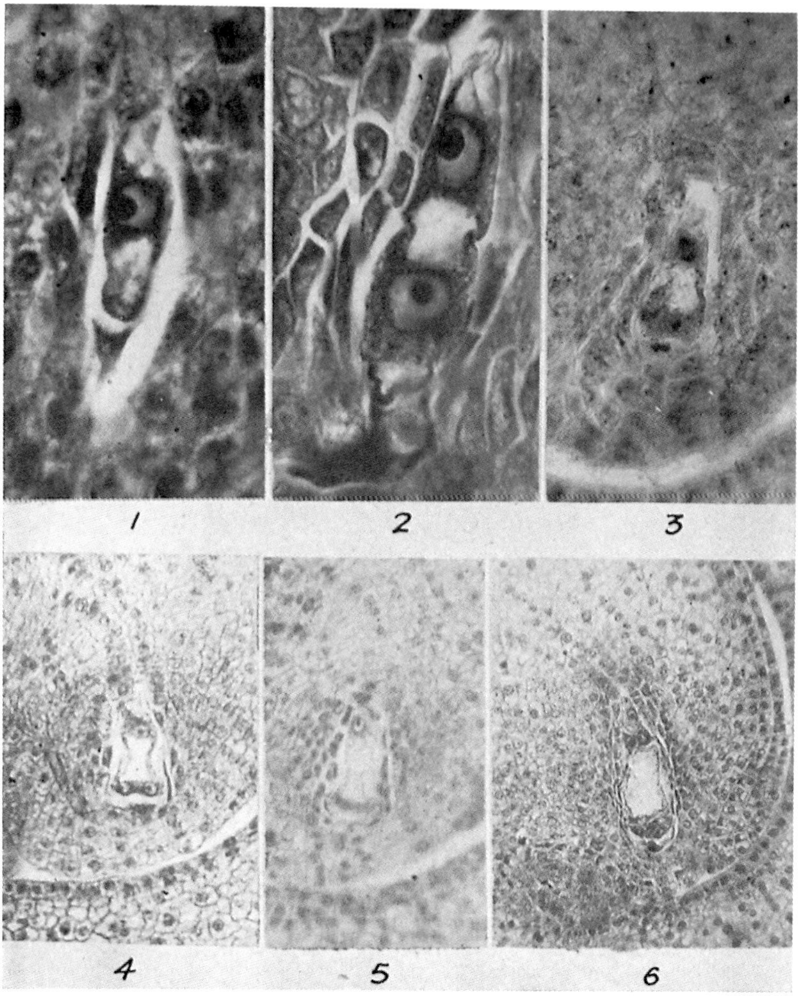

FIG. 43.—Photomicrographs. Vertical sections through nucellus showing development of embryo sac after reduction-division shown in Figure 42. (Figure 41 shows that the embryo sac gradually becomes nearly basal within anterior side of nucellus.)

1, one-neucleate stage of embryo sac, being the persisting megaspore. X 1000.

2, two-nucleate stage of embryo sac, showing disintegrating remnants of three outer spores (complete spikelet shown in Figure 41—4). X 1200.

3, mitosis in the formation of the four-nucleate embryo sac. X 500.

4 and 5, two consecutive microtome sections of an embryo sac in the four-nucleate stage (three nuclei in one section and one in the other). X 500.

6, eight-nucleate stage of embryo sac before central migration of the two polar nuclei which join in formation of the primary endosperm nucleus upon fertilization. (Complete spikelet shown in Figure 41—5.) X 300.

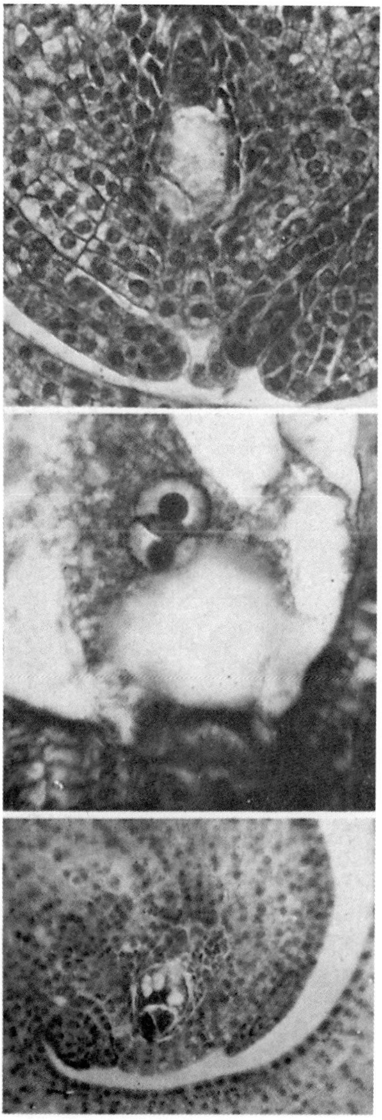

FIG. 44.—Photomicrographs. The mature embryo sac before fertilization, shown in three consecutive microtome sections.

Lower, egg and two synergids. X 300.

Middle, two polar nuclei. X 1100.

Upper, three antipodal cells. X 450.

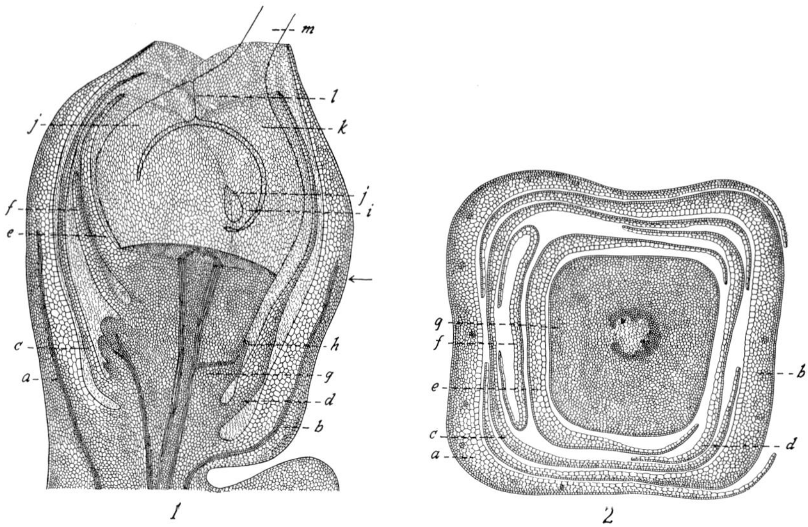

FIG. 45.—Semi-diagrammatic reconstruction of the pistillate spikelet to show the arrangement of the various organs at time of silking.

1, longitudinal section; 2, cross section (arrow in "1" indicates point of cross section). X 30.

a, first or outer glume; b, second or inner glume; c, lemma of rudimentary flower; d, lemma of functional flower; e, palet of functional flower; f, palet of rudimentary flower; g, axis of functional flower showing vascular system; h, rudimentary stamen; i, nucellus; j, embryo sac; k, carpels or ovary wall; l, stylar canal; m, silk.

of the developing embryo sac. The two nuclei, by two successive divisions (Fig. 43), form a group of four nuclei at each end of the embryo sac.

One nucleus from each group next moves toward the center of the embryo sac until they meet and remain in contact (Fig. 44–middle) a little above the egg cell, but do not fuse until after fertilization. The three nuclei at the chalazal end, farthest from the micropyle, divide until a group of about 20 to 40 cells is formed. This division of the antipodal cells is characteristic of the grass family.

One of the three nuclei at the micropylar end enlarges and becomes the nucleus of the egg, while the others become the nuclei of the synergids (Fig. 44–lower). The embryo sac now enlarges, and without further nuclear divisions becomes ready for fertilization. The fully organized embryo sac then has at one end the large egg cell with the two synergids beside it. A little nearer the middle are the two polar nuclei, and at the opposite end is the group of antipodal cells, many of which have more than one nucleus *(18* and *65)*.

The embryo sac is now ready for fertilization but if pollination is prevented it may remain in this condition for some time, perhaps two weeks, after which the embryo sac and nucellus disorganize and fertilization is no longer possible. For the sake of clarity, the structure of the pistillate spikelet

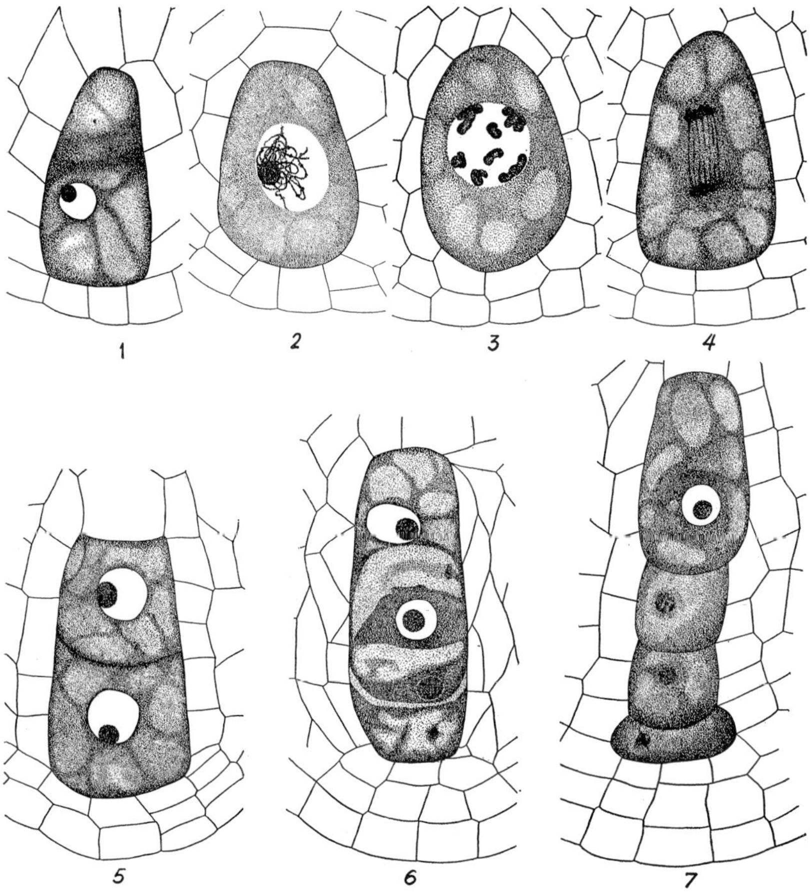

Fig. 46.—Longitudinal sections through part of nucellus before and during megasporogenesis. During this period and that of embryo-sac development, the nucellar epidermis increases from one to five or six cells in thickness by periclinal division. X 600.

1, archesporium, which functions directly as the megaspore mother cell since no wall cell is cut off as in the development of the microspore shown in Figures 28 and 29.

2-5, stages during first meiotic division (2, zygonema; 3, diakinesis; 4, telephase; 5, two-cell stage).

6-7, quartet or four-spore stage formed by second meiotic division, showing progressive degeneration of three outermost spores, and persistence of the innermost spore which normally functions as the embryo sac.

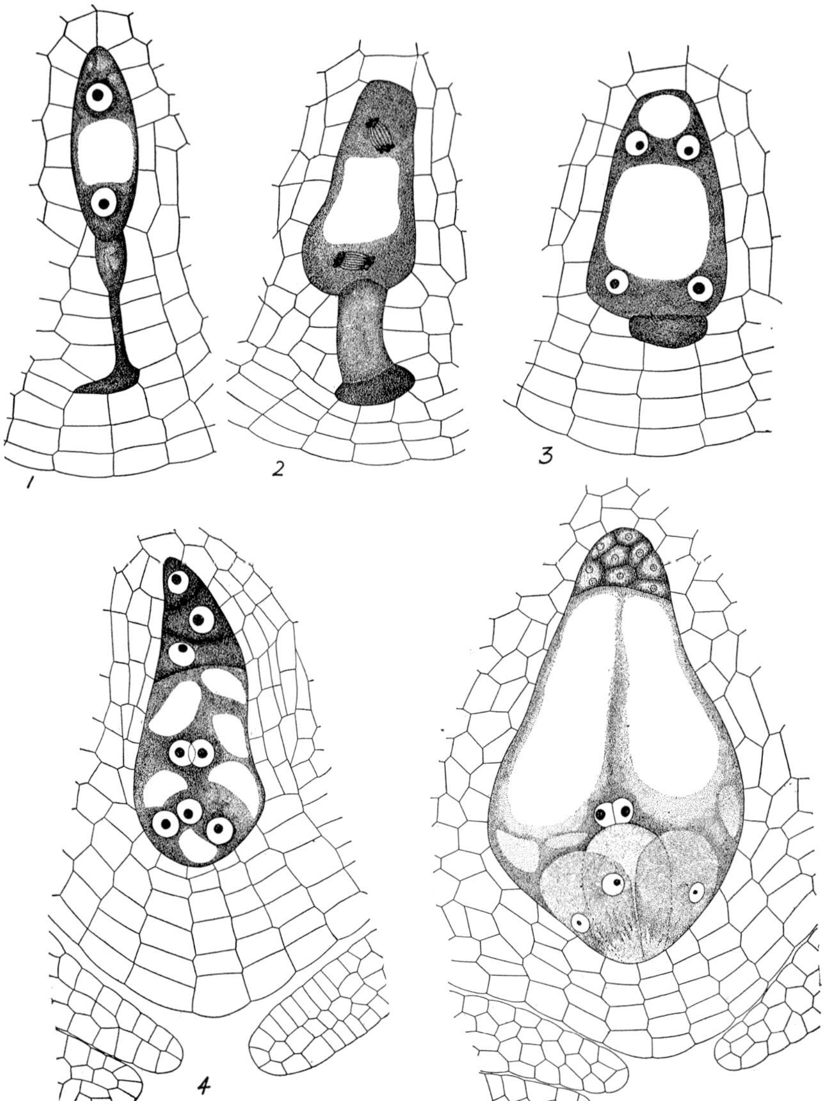

FIG. 47.—Successive stages in the development of the embryo sac.

1, two-nucleate stage of embryo sac, dark mass below being the remnant of the three non-functioning megaspores. X 500.

2, two nuclei dividing to form four. X 500.

3, four-nucleate stage, last trace of three degenerating spores showing below. X 500.

4, eight-nucleate stage; antipodal cells formed above; polar nuclei have approached each other near center; egg and synergid nuclei below. X 400.

5, mature embryo sac; egg and synergids organized as cells; polar nuclei in contact above egg; antipodals have divided to form a group of cells, some of which are binucleate. Vacuoles have formed in cytoplasm surrounding polar nuclei. X 250.

and the stages of meiosis and embryo-sac development are reconstructed by drawings in Figures 45 to 47. An ear shoot with husks partly removed, at the fertilization stage, and two other ear shoots illustrating the effect of fertilization on growth of the silk are shown in Figure 48. Sections of both normal and abnormal silks are shown in Figure 49 as evidence of the three-carpel origin of the grass pistil.

REPRODUCTION AND KERNEL DEVELOPMENT

Pollination and Germination of Pollen

WHEN POLLEN is shed by the tassels, only that which is intercepted by fresh silks can germinate, and where several pollen grains germinate on the

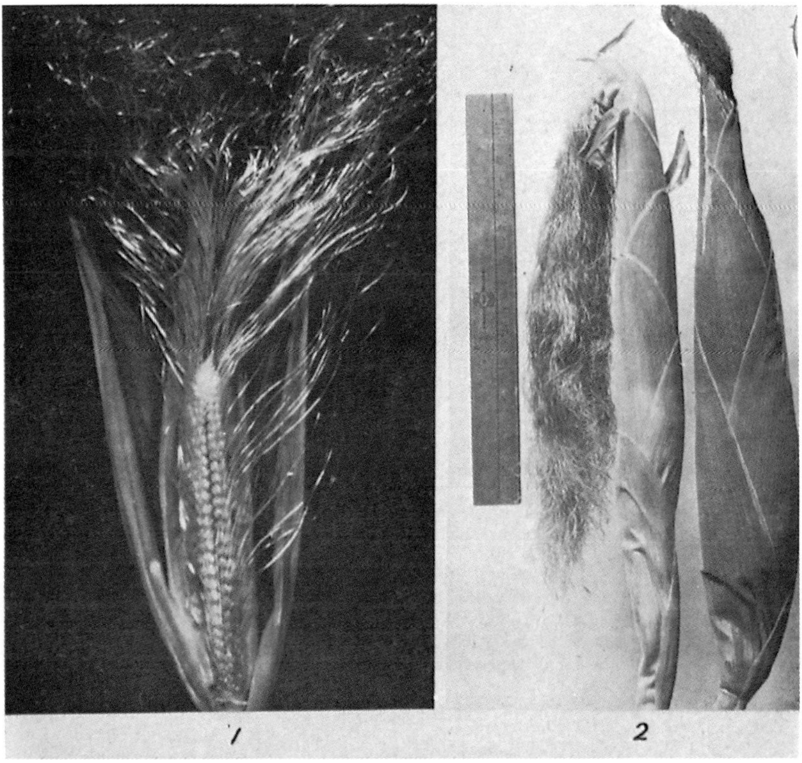

FIG. 48.—Photographs.

1, an ear shoot with about 800 pistillate flowers, ready for fertilization (husks partly removed). X 0.7.

2, ear shoots showing the effects of fertilization on growth of the silks. X 0.2.

Left, pollen was withheld by enclosing the ear shoot in a paper bag before silk emergence. This resulted in continued growth of the silks for a period of 12 days to an exposed length of 15 inches. The fresh silk is receptive to pollen throughout its entire length.

Right, silks were pollinated soon after their emergence and further growth ceased within a day thereafter.

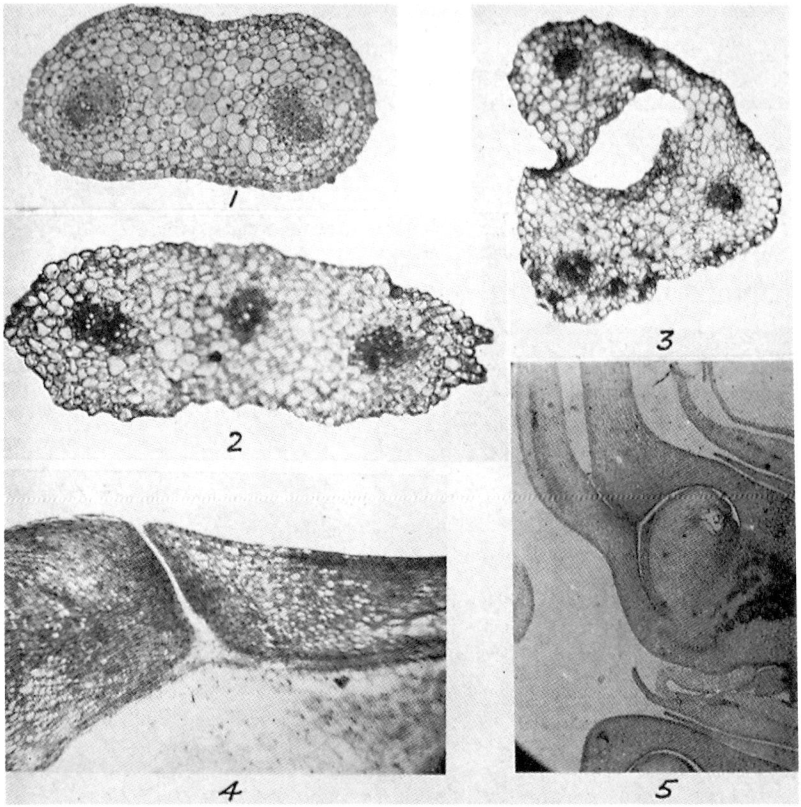

FIG. 49.—Photomicrographs. The silk, with evidence of tricarpellary origin.

1, cross section of a normal style showing two vascular strands, one from each anterior carpel. X 125.

2, cross section of an occasional style in which the posterior carpel also has participated in the formation of the silk, supplying it with three vascular strands. X 125.

3, cross section of an occasional hollow silk which is caused by extension of the stylar canal into the interior of the silk when all three carpels join in its formation. X 100.

4, normal stylar canal located where the two anterior and the one posterior carpels join. X 120.

5, longitudinal section of pistil with a tricarpellary, hollow silk. (Compare with Figures 41—4 and 5.) X 40.

same silk usually only one functions in fertilization. Most of the pollen is lost by falling to the ground or is caught by the leaves and accumulates in the leaf axils. In windy weather part of the pollen may be blown out of the cornfield.

In pollination, the pollen is usually caught by the hairs of the silk (Figs. 50 and 51) although it may function when caught on the body of the silk *(21)*. The silk supplies the pollen with moisture which causes it to germinate, sending out a pollen tube through the germ pore (Fig. 33). The pollen tube may follow the surface of the hair for some distance but usually soon enters between the cells of the hair from which it enters the silk proper (Figs.

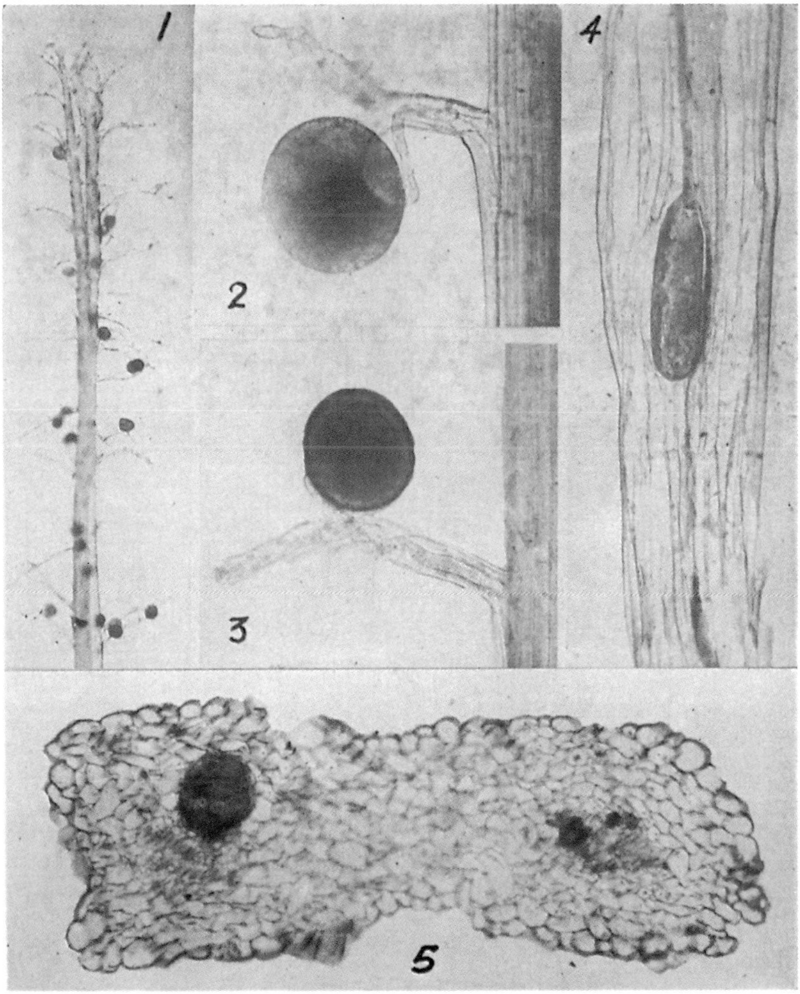

FIG. 50.—Photomicrographs showing pollination of the silk and growth of the pollen tube.
1, exposed tip of silk with pollen clinging to its hairs. X 18.
2, germinating pollen with tube starting down the silk hair. X 200.
3, pollen grain with tube grown full length of hair and into body of silk. X 160.
4, longitudinal section of pollen tube with its enlarged head growing down through body of silk. X 200.
5, cross section of silk showing head of one pollen tube at left following the conducting tissue adjacent to the vascular strand and the tails of three pollen tubes in cross section at the right. X 250.

50 and 51). Here it soon reaches and follows the sheath cells surrounding the vascular tissue to the base of the silk. At this point the conducting tissue leaves the bundle and passes to the cavity of the ovary. On the inner wall of the ovary the strand of conducting tissue turns downward and may extend on the inner surface of the carpels to the micropyle. It was actually traced, however only a short distance on the inner surface of the carpels

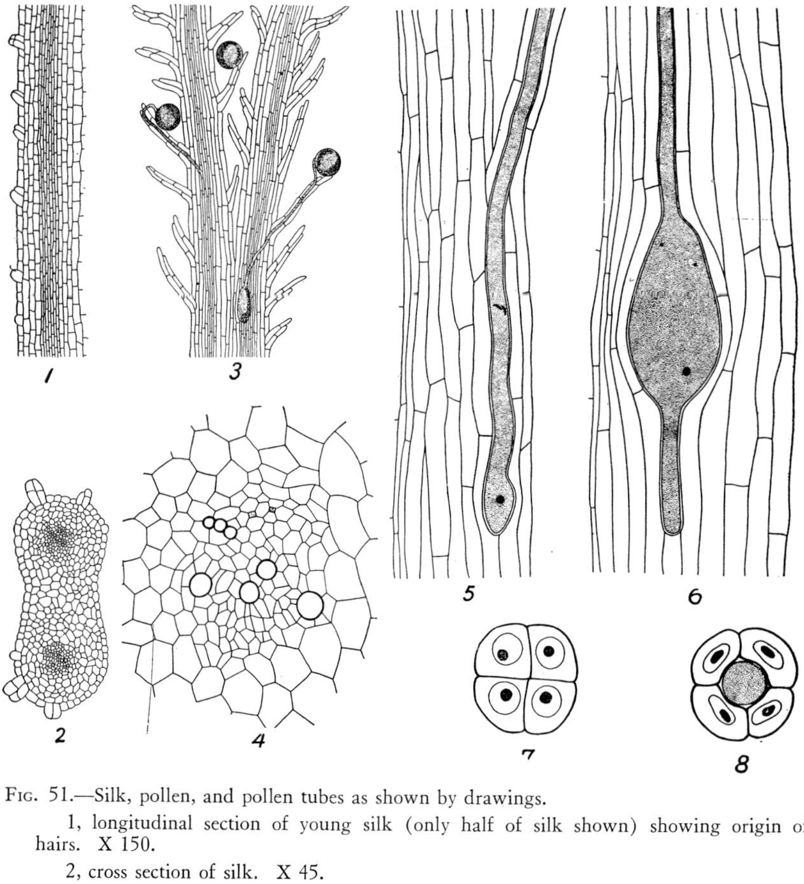

FIG. 51.—Silk, pollen, and pollen tubes as shown by drawings.

1, longitudinal section of young silk (only half of silk shown) showing origin of hairs. X 150.

2, cross section of silk. X 45.

3, longitudinal section showing tip of silk with hairs, pollen, and germinating pollen. X 45.

4, cross section of vascular bundle and conducting tissue of silk. Three vessels and sieve tubes with companion cells above and four pollen tubes below. X 420.

5, pollen tube with tube nucleus and two sperms just entering body of silk from hair. X 270.

6, pollen tube with tube nucleus and two sperms in typical development in the body of the silk. X 270.

7, cross section of hair showing cell walls and nuclei. X 750.

8, cross section of hair with pollen tube in center. X 750.

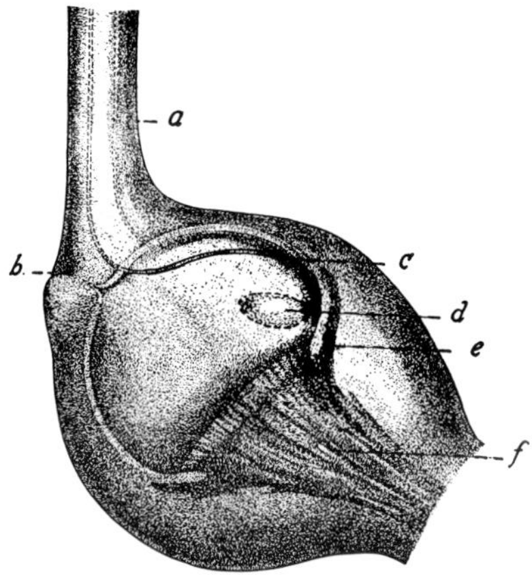

Fig. 52.—Diagrammatic drawing of pistil (transparency) showing path of pollen tube from silk to embryo sac. X 40.

a, silk; b, stylar canal; c, pollen tube; d, embryo sac; e, one of two vascular bundles extending through silk; f, vascular tissue.

(65 and *75)*. The structure of the normal silk, silk hair, and growing pollen tube is shown in Figure 51.

Fertilization

The path of the pollen tube from the silk to the embryo sac is shown diagrammatically in Figure 52. When the tip of the pollen tube reaches the micropyle it grows between the protruding cells of nucellar tissue until the embryo sac is reached. On entering the embryo sac (Figs. 53 and 54) the end of the pollen tube ruptures, setting free the two sperms. The nucleus of one sperm fuses with the egg nucleus, forming the zygote within which the chromatin from both sources completely intermingles *(75)*. This restores the diploid number of 20 chromosomes to the somatic cells of the new sporophyte represented by the zygote or first cell of the proembryo from which the embryo develops.

The other sperm nucleus fuses with one of the two polar nuclei *(75)* and this fused nucleus in turn fuses with the other polar nucleus, thereby establishing the primary endosperm nucleus with 30 chromosomes. This double fertilization explains how characters of the male parent may show up in the endosperm as well as in the embryo and the new plant into which it will develop. The mature embryo sac undergoing double fertilization, and initial stages of the embryo and endosperm are illustrated by drawings in Figure 54.

Development of the Endosperm

The primary endosperm nucleus, resulting from fusion of a sperm with the two polar nuclei (Fig. 54–2 and 3), divides mitotically within 3 to 5 hours after fertilization, and repeated divisions continue until a number of free nuclei (Fig. 54–4) are formed. The embryo sac enlarges and a central vacuole is formed, surrounded by a layer of cytoplasm in which are embedded the free nuclei calculated to number about 128 to 256, 50 hours after pollination. Cell walls are formed between the nuclei but no wall is formed next to the central vacuole; each free nucleus divides and a cell wall is formed between the resultant two so that one becomes enclosed in cell walls and the other remains free on the side toward the central vacuole (Fig. 54–5). This process is repeated until the cavity becomes filled with cellular tissue. These early stages in the development of the endosperm and embryo are also shown by photomicrographs in Figures 55 and 56 (right).

Shortly after the vacuole of the endosperm has become fully occupied by cells, the cells of its outer layer just over the placenta become differentiated (Fig. 57) and function as a food-conducting tissue. This differentiation continues with growth until it extends over the entire region of placentation.

This specialized basal tissue of the endosperm was shown in drawings of the corn kernel without explanation by Harz *(33)*. The corresponding region in Johnson grass was called compressed and empty endosperm cells by Harrington and Crocker *(32)*; and in the seed of sugar cane they are called empty

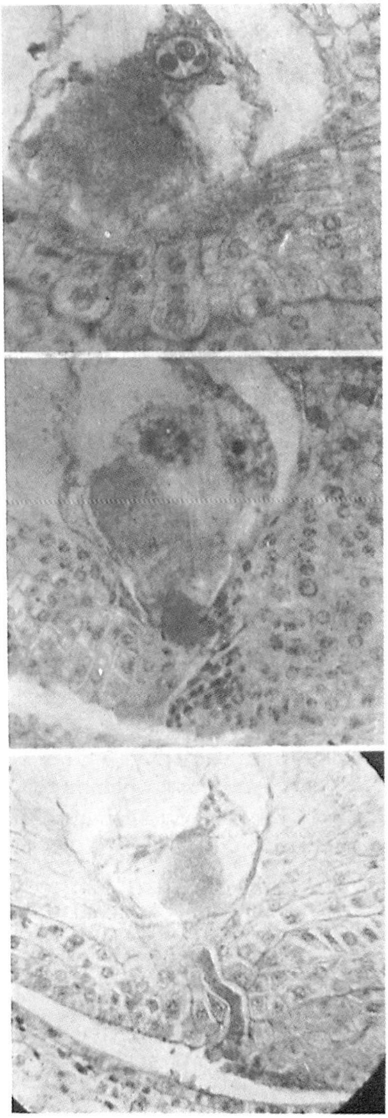

FIG. 53.—Photomicrographs. Double fertilization shown in three consecutive microtome sections. (Compare with Figure 44.)

Lower, pollen tube enters embryo sac through micropyle and intervening nucellar tissue, ruptures, and releases its contents, including the two sperms. X 450.

Middle, at right, egg and sperm about to unite to form zygote. X 500.

Upper, the two polar nuclei with a sperm nucleus fused with one of them just prior to fusion with the other polar nucleus in formation of the primary endosperm nucleus. X 650.

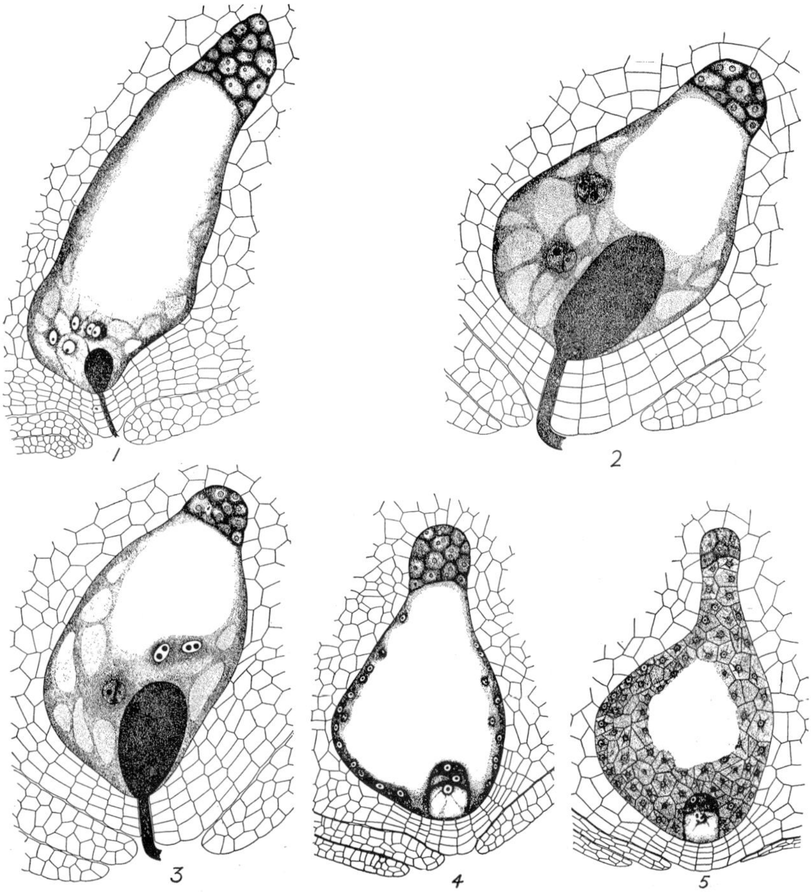

FIG. 54.—Double fertilization and young embryo and endosperm stages.

1, pollen tube just entered into embryo sac (i.e., female gametophyte) and contents discharged, the sperms still within the cytoplasm from tube (i.e., in the cytoplasm of the male gametophyte). X 120.

2, after fertilization, showing one male (small) nucleus fusing with the egg nucleus to form the zygote, and the other fusing with one of the two polar nuclei preliminary to the triple fusion to form the primary endosperm nucleus. X 160.

3, endosperm nucleus divided, 27 hours after pollination (approximately 3 hours after fertilization). Many of the endosperm nuclei have two nucleoli. X 160.

4, three-cell stage of embryo, and free-nuclear stage of endosperm, 40 hours after pollination. X 80.

5, slightly older stage of embryo, and endosperm becoming cellular, 100 hours after pollination. The antipodals, many of them binucleate, have about reached their final number. X 60.

elongated endosperm cells by Artschwager, Brandes, and Starrett *(5)*. In Coix and other plants Weatherwax *(96)* calls them placental tissue of the endosperm.

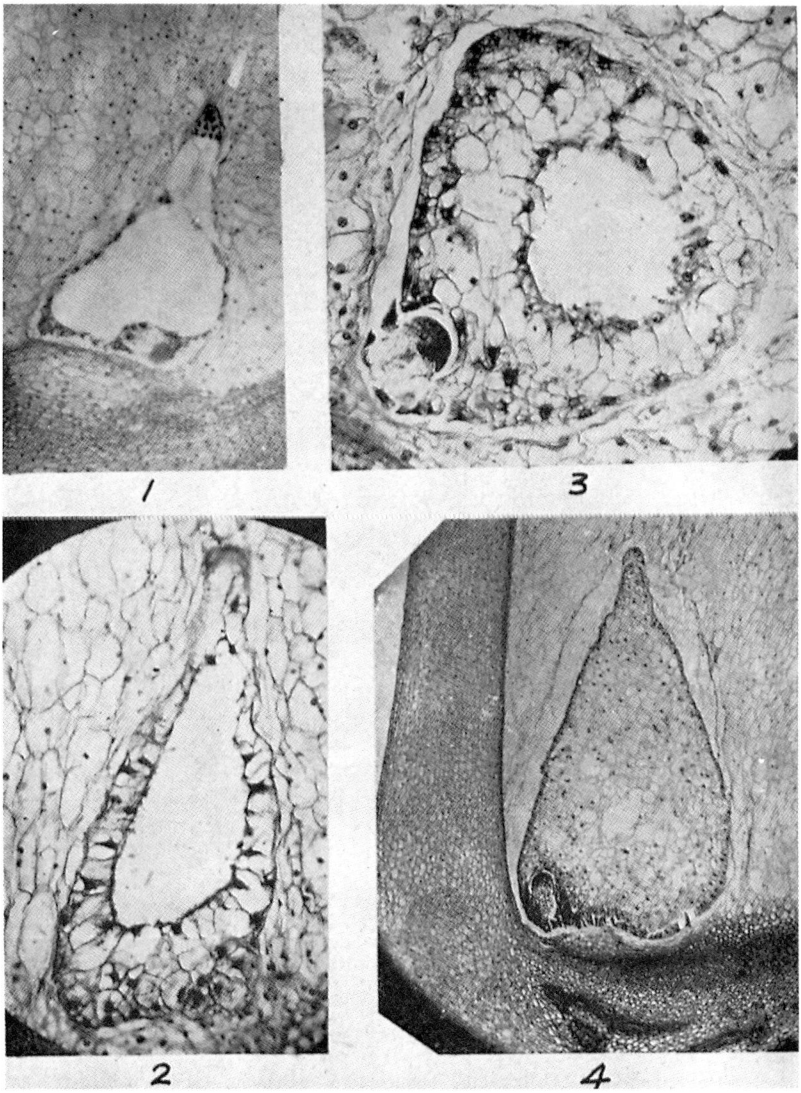

FIG. 55.—Photomicrographs. Early stages of endosperm, also embryo.

1, longitudinal section of enlarged embryo sac showing free endosperm nuclei lining its cavity, antipodal cells at its apex and the proembryo at its base, 36 hours after pollination. X 125.

2, endosperm becoming cellular. X 125.

3, the hollow endosperm is filling up centrally by formation of new cells laid down by division of the free nuclei which are readily visible, 100 hours after pollination. X 100.

4, the endosperm becomes solid and digests away the surrounding nucellar tissue which it replaces during growth. Basal cells of endosperm start differentiation at this stage, 145 hours after pollination. X 50.

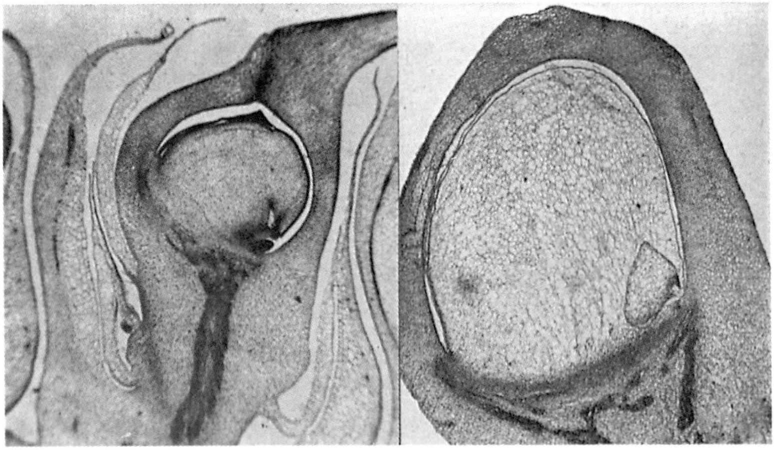

FIG. 56.—Photomicrographs. Complete ovaries showing transformation of ovule with mature embryo sac just before fertilization (left, X 36) to one with developing embryo and endosperm four days after fertilization (right, X 32). The brown hilar tissue of mature seeds (shown forming in Figures 57 and 62—4) develops in the region of relatively small parenchyma cells just below the ovule, and extends entirely across the plane of ovule attachment within the ovary wall.

REGION OF CELL DIVISION

For some time after the endosperm becomes cellular, cell divisions occur throughout (Fig. 58) but soon cease or at least become rare in the interior, divisions then being confined to cells of the peripheral zone, several cells deep. This zone, except over the hilar region, is meristematic and functions much like a cambium, as was reported by Gordon *(30)* in several of the cereals and more recently by Fisk *(27)* and Randolph *(71)* in corn. No cell divisions were noticed within this lower part of the endosperm though a careful search was made for mitotic figures there, as according to Lampe *(50)* this should be where growth takes place. The young cells formed inside the layer of meristem may continue to divide for a short time before differentiating into regular endosperm cells. Until about 20 days after pollination, the outermost layer acts as part of the cambium-like meristem and periclinal cell division within it is of frequent occurrence. Thereafter its growth occurs only by anticlinal division and enlargement of cells. Later these surface cells become differentiated into the aleurone layer.

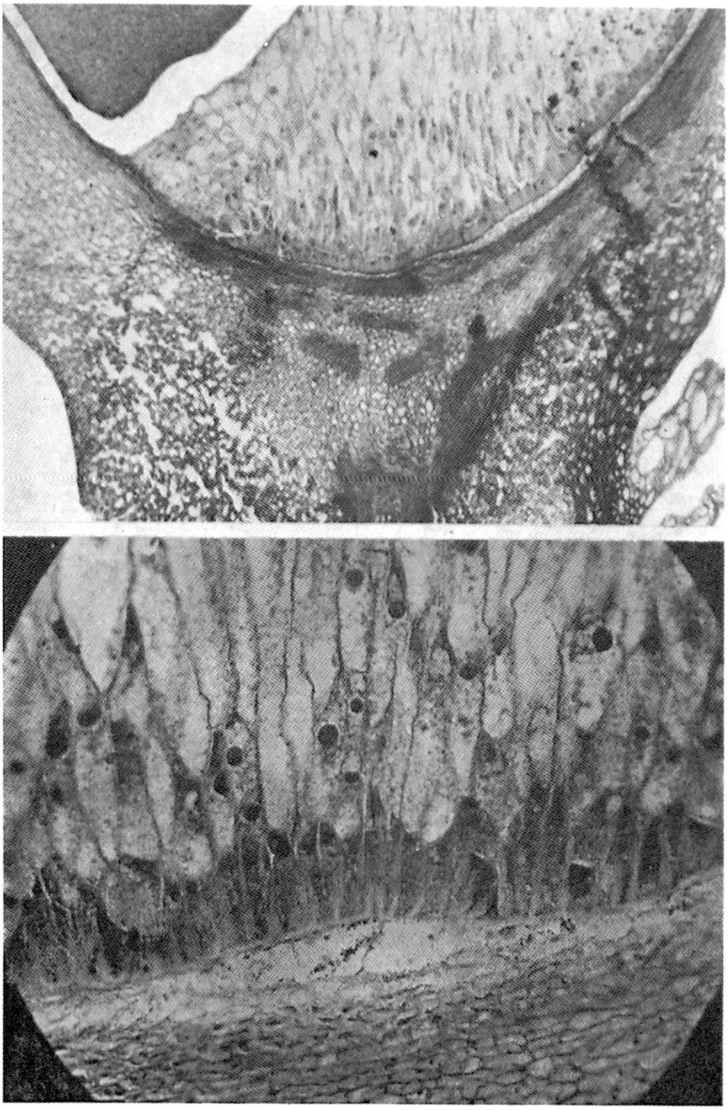

FIG. 57.—Photomicrographs. Specialized conducting cells of endosperm.

Above, base of kernel showing specialized conducting cells of endosperm, and also the abscission layer over the region of placentation. X 30.

Below, conducting cells in lower portion of endosperm; also abscission tissue in ovary wall. X 120.

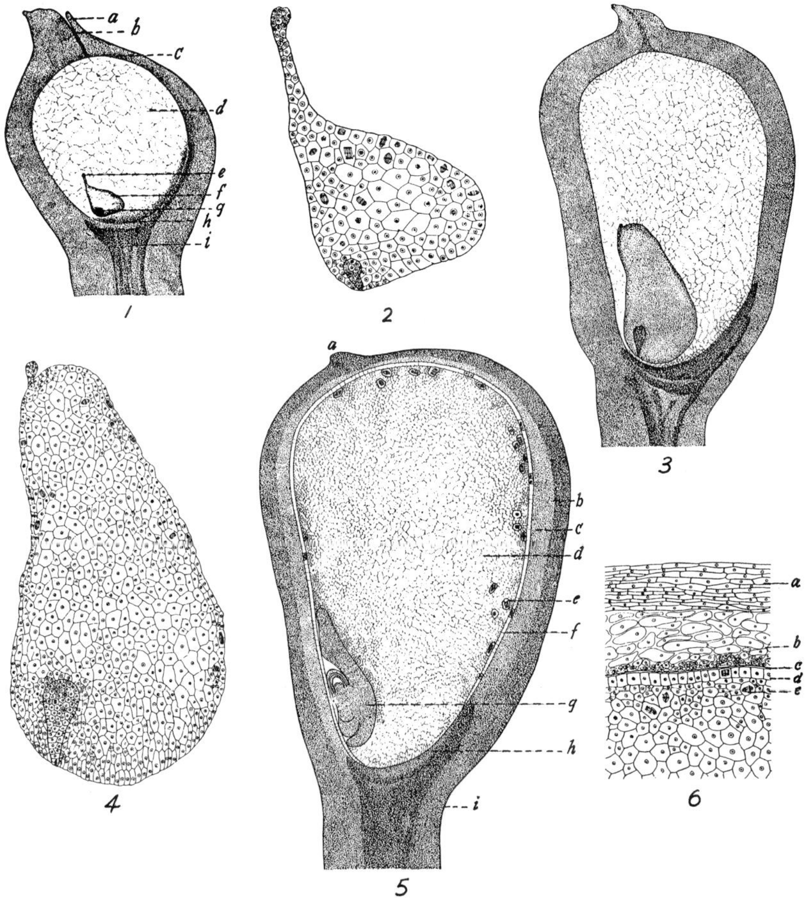

FIG. 58.—Series of longitudinal sectional drawings of developing kernel after fertilization to show region of cell division in endosperm.

1, pistil, five days after fertilization. a, silk scar; b, stylar canal; c, carpels which will form pericarp; d, nucellus; e, antipodals; f, endosperm; g, embryo; h, placental-hilar-funicular region; i, vascular tissue of carpels. X 10.

2, endosperm, embryo and antipodals from "1." Cell division as indicated by mitosis, generally scattered throughout endosperm up to this stage. X 55.

3, kernel about 10 days after pollination. X 10.

4, endosperm, embryo, and antipodals from "3." Cell division has become limited to outer region of endosperm, away from the embryo. No cell division was ever found in basal part of endosperm except in the youngest stages. X 30.

5, kernel about 20 days after pollination. a, silk scar; b, pericarp; c, inner part of carpels breaking down; d, endosperm; e, location of mitosis in endosperm; f, aleurone; g, embryo; h, hilar region; i, pedicel. X 10.

6, section through endosperm and pericarp of "5" to show detail. a, pericarp; b, inner part of carpels breaking down; c, nucellar membrane; d, aleurone; e, endosperm showing mitosis. X 30.

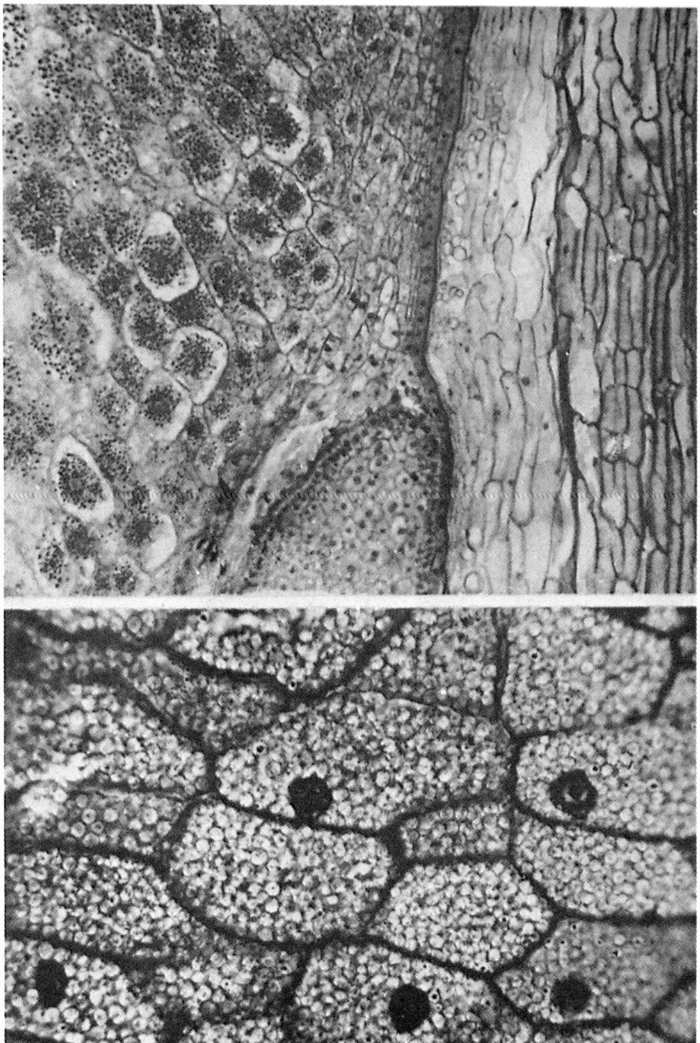

FIG. 59.—Photomicrographs. Starch grains in endosperm.

Above, longitudinal section of portion of growing kernel, 25 days after pollination, showing at left: endosperm with first starch granules, smaller cells of endosperm near pericarp where cell division and growth take place; at right: pericarp; below: tip of embryo. X 120.

Below, mature endosperm cells packed with starch grains. X 200.

STARCH FORMATION

All the cells of the endosperm except the surface layer serve for storage of food materials, consisting largely of starch grains in field corn. The aleurone layer has often been considered as a food-storing region but it seems well established that it furnishes enzymes for digestion and is not itself used up. The cells of the aleurone layer appear intact and still filled after the food reserves of the rest of the endosperm are completely exhausted during seedling growth. Starch formation begins two weeks or less after fertilization. The first cells to show starch grains are in the upper or crown part of the kernel and this formation progresses toward the basal part of the endosperm. There will thus be the most starch in the cells of the upper part of the endosperm and gradually less toward the base, forming a major gradient in starch formation.

As long as new cells are forming near the surface these will contain smaller starch grains and less starch than the cells farther from the surface. There will then be two gradients of starch formation, a major one in which the cells become richer in starch content in passing upward from the base of the kernel toward the crown *(50)*; and a minor starch gradient in passing from the youngest cells near the surface toward the interior (Fig. 59).

VARIATIONS IN ENDOSPERM

In ordinary field corn the accumulation of starch continues until all the interior cells contain large quantities (Fig. 59), but the cells are not filled

FIG. 60.—Types of corn differing in character of endosperm.
Left to right: heterozygous pod corn which may have any type of endosperm, pop corn, flint corn, dent corn, soft (flour) corn, and sweet corn.

equally in all regions of the endosperm. In the "horny" region the starch grains are closely packed and angular and the spaces are filled with protein, whereas in the "starchy" region the starch grains are rounded and less closely packed. Dent and flint corn differ in the amount of horny starch in the upper part or crown of the kernel; in the flint and pop types this horny layer extends over the top of the kernel so there is no denting due to shrinkage as in dent corn (Fig. 60). Soft (flour) corn has no horny endosperm and therefore also does not dent. Waxy corn differs from other starchy corn in that the starch of its endosperm consists entirely of amylose rather than containing also amylopectin. It stains reddish-brown instead of blue when treated with iodine. In sweet corn there is less starch and more sugar and water-soluble polysaccharide, causing the whole kernel to become wrinkled in drying.

At first the endosperm in all corn varieties is white, but in about two weeks after fertilization the endosperm and aleurone colors characteristic of the particular variety appear. In yellow corn this color is diffused throughout the endosperm. In colored aleurone, either red or purple, the coloring is limited to this layer. Most red corn owes its color to a red pericarp, which also shows up in about two weeks. Sun red, also a pericarp color, appears only in kernels exposed to sunlight. In some varieties a red pericarp is present over a purple aleurone, causing the kernel to appear very dark red.

The Antipodal Cells in Later Stages of Kernel Development

The antipodals in corn were assumed to degenerate and disappear during the early stages of the development of the kernel, as is true in most other grasses, until Weatherwax *(95)* reported finding their remnants in kernels nearly or quite mature. In these studies also, what appears to be such remnants were sometimes found in late stages of kernel development. However, in many sections they appeared to be degenerating shortly after the central vacuole of the endosperm had been filled with cells. Randolph *(71)* has also reported finding the antipodals persisting in late stages of kernel development and such persistence was common in the corn examined. He suggests that the persistent antipodal cells become rather similar in appearance to the cells of the endosperm. Their persistence in corn appears to be rather exceptional.

Development of the Embryo

The fertilized egg does not begin to divide as soon after fertilization as does the primary endosperm nucleus, and the young embryo or proembryo develops much more slowly than the endosperm in its early stages. At the first division of the fertilized egg about 40 hours after pollination the endosperm may have from 8 to 32 free nuclei and by the time wall formation begins in the endosperm about four days after pollination, the proembryo commonly does not have more than about a dozen cells while the endosperm will have 250 or more.

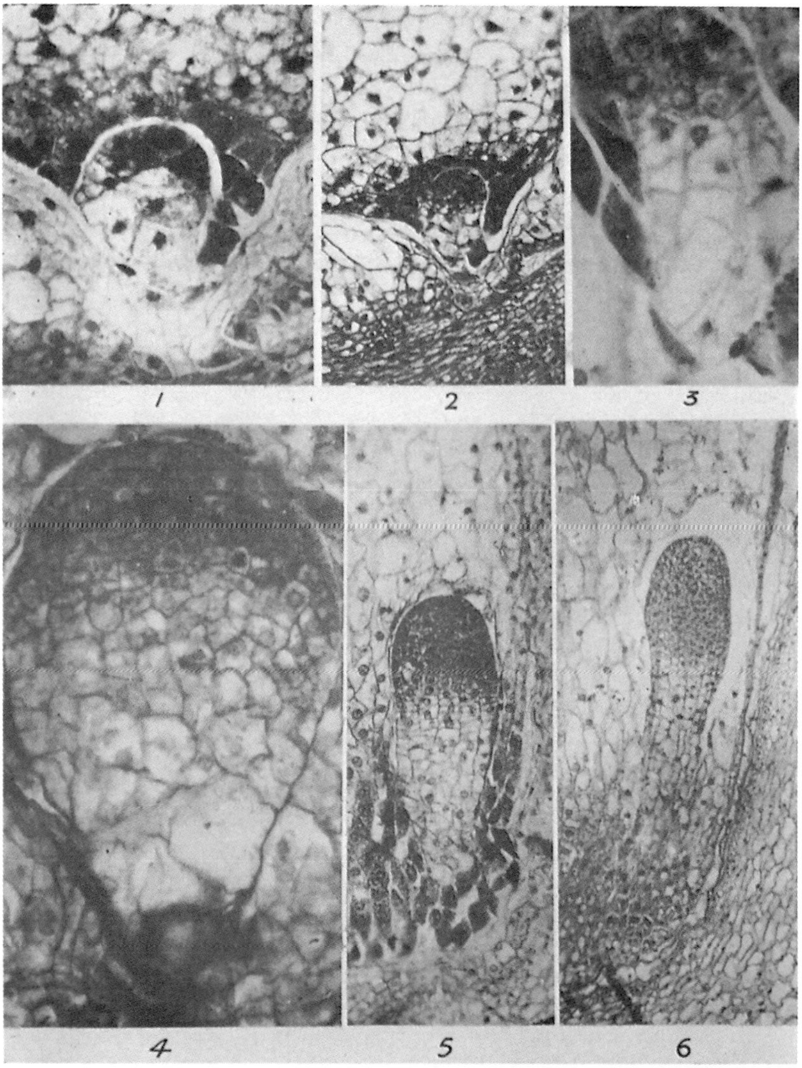

Fig. 61.—Photomicrographs. Longitudinal sections showing growth of the proembryo and differentiation of embryo at its tip (various magnifications). Because of the more rapid apical growth, the embryo becomes club shaped. Prior to the ninth day after pollination there is no cellular differentiation and peripheral cell division is both periclinal and anticlinal. By the ninth day, the epidermis becomes differentiated apically and the cell division within it thereafter is chiefly anticlinal.

1, four days after pollination, X 400; 2, five days, X 180; 3, six days, X 400; 4, eight days, X 400; 5, nine days, X 180; 6, ten days, X 160.

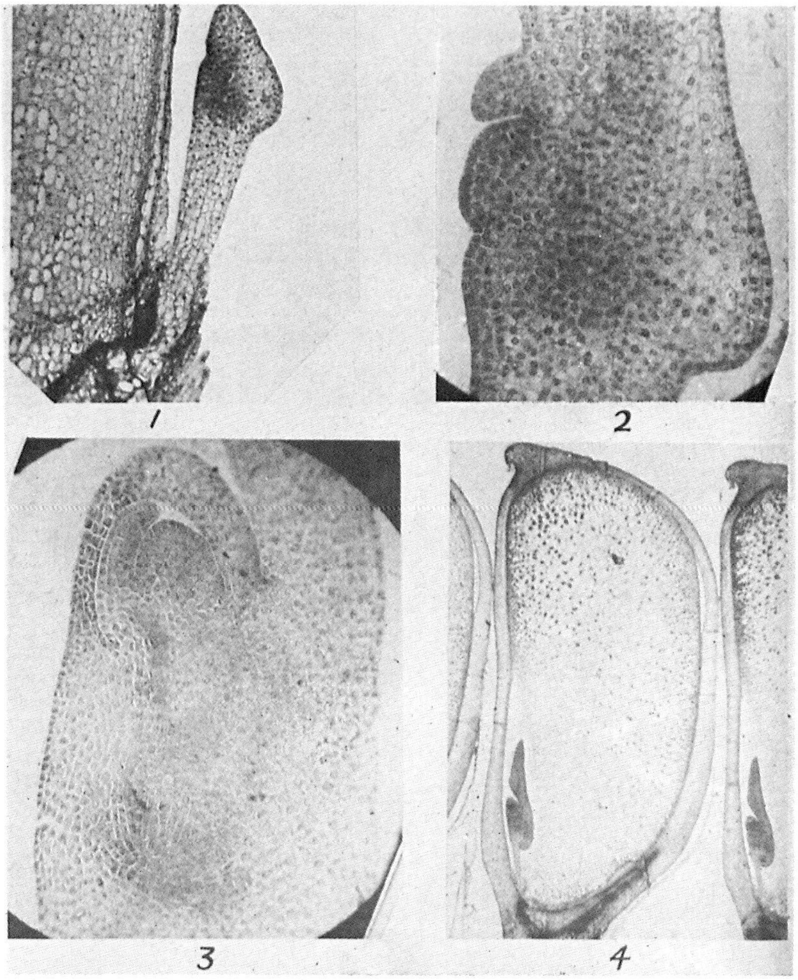

FIG. 62.—Photomicrographs. Longitudinal sections showing differentiation of organs of the embryo at successive intervals after pollination.

1, thirteen days after pollination. Elongated suspensor of embryo is surmounted by scutellum initial above and at right, and shows first evidence of the growing tip of plumule at slight bulge below notch at left. The differentiating plumule-radicle meristem is represented by the more deeply stained region. The wide tissue at left of embryo is the ovary wall (endosperm is removed). X 120.

2, fifteen days. Slightly more advanced stage than "1," showing the stem tip, at left, with the coleoptile growing out as a ridge around it. The region of radicle differentiation is apparent. Scutellum above and at right. X 150.

3, twenty-one days. The plumule-radicle axis is well differentiated. Stem tip and first leaf of plumule enclosed in coleoptile, above, and radicle and coleorhiza initials below. X 80.

4, twenty-one days. Embryo, endosperm, and pericarp showing silk attachment. At this stage the stem tip clearly appears terminal. Seed carefully cured at this age may germinate.

PROEMBRYO

The early stages of development (Fig. 61) show such an irregular arrangement of cells that it is not possible to trace the future organs of the embryo back to definite cells or groups of cells in the proembryo as can be done in some plants. The basal cells form the suspensor whose growth serves to orient the embryo properly with respect to the endosperm. The suspensor soon ceases to enlarge and degenerates, although remnants of it may remain at the base of the embryo for a long time. Food reaches the embryo by way of the endosperm and scutellum.

The proembryo soon becomes a club-shaped mass of cells, showing little differentiation except that the lower part, the suspensor, consists of larger cells with less cytoplasm than does the upper part in which the cells are small and densely filled with protoplasm.

The first sign of further differentiation of the embryo is the appearance of a region of small cells that are more densely filled with protoplasm on the anterior side of the embryo a little below its tip. The stem tip develops from this region and soon becomes evident by the formation of a protuberance with a notch above it (Fig. 62).

A common explanation has been that the scutellum or cotyledon is terminal, and the stem tip lateral. It seems more likely, however, that the

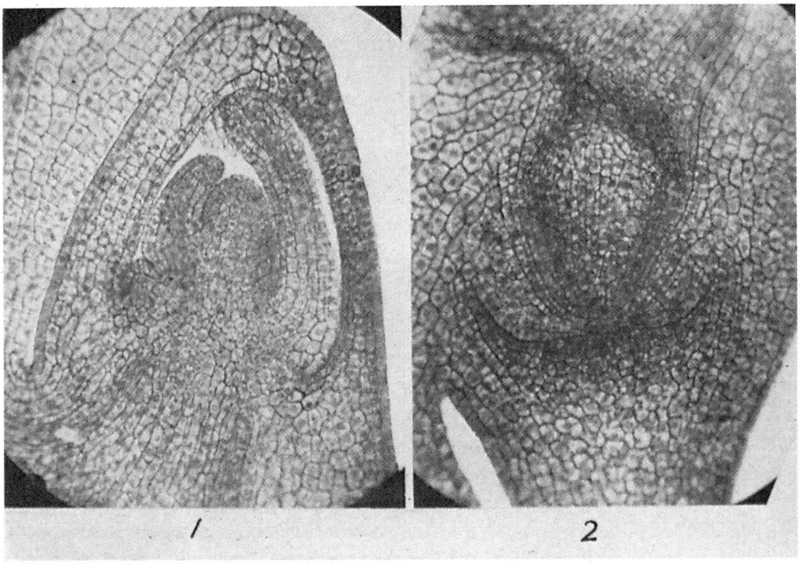

FIG. 63.—Photomicrographs. Longitudinal sections of clearly differentiated plumule and root initials of embryo 25 days after pollination.
1, plumule with two leaf initials enclosed by coleoptile. X 140.
2, primary root initial enclosed in the coleorhiza. X 140.

rapid development of the tissues of the scutellum pushes the original tip of the proembryo aside, making the stem tip appear lateral in early development. Rapid growth of the leaves as compared with that of the stem is a conspicuous characteristic of grasses as it is of many other monocotyledons.

DEVELOPMENT OF PLUMULE AND SEMINAL ROOTS

A ring of tissue forms around the stem tip which develops into the coleoptile (Figs. 62 to 64). There has been much dispute over the homology of the coleoptile. Some have considered it as an independent leaf, but this would leave two successive leaves, the scutellum and coleoptile, on the same side of the stem instead of alternating as do all the other leaves. The other view is that it is an outgrowth of the scutellum and that the two together make up the cotyledon. In this paper, both are regarded as distinct modified leaves—the first two of the plant—and separated by the first internode.

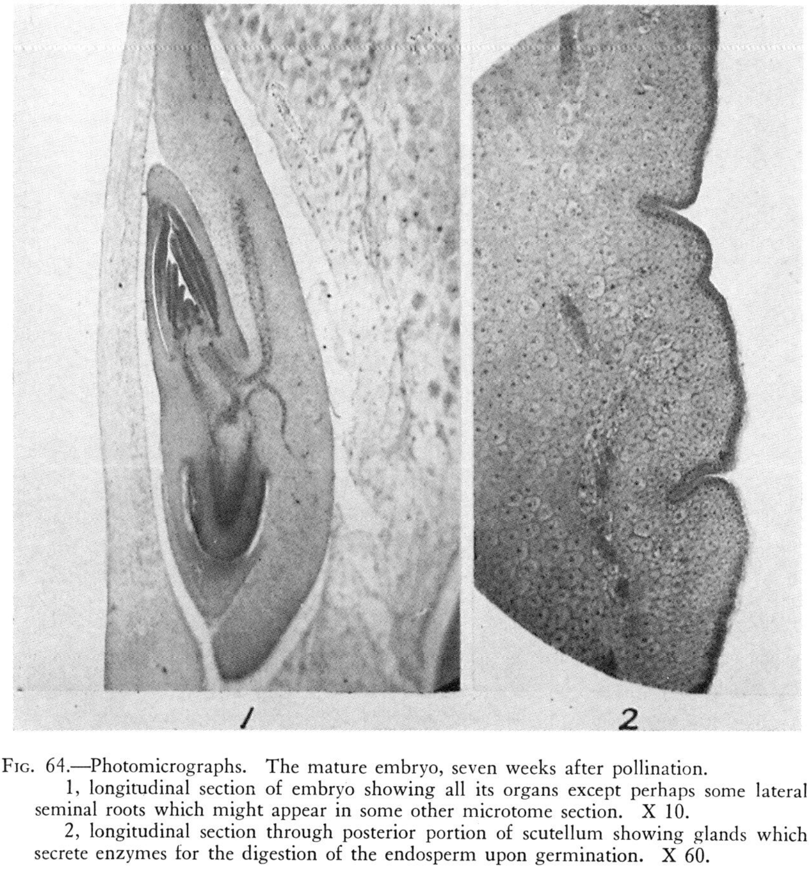

FIG. 64.—Photomicrographs. The mature embryo, seven weeks after pollination.

1, longitudinal section of embryo showing all its organs except perhaps some lateral seminal roots which might appear in some other microtome section. X 10.

2, longitudinal section through posterior portion of scutellum showing glands which secrete enzymes for the digestion of the endosperm upon germination. X 60.

Structurally the coleoptile is quite different from the true leaves of the plumule, in some respects resembling a prophyllum, as for instance in having two main veins corresponding to those of the keels of a prophyllum, and no midrib.

If one accepts the explanation that the coleoptile is part of the cotyledon, then in corn as in many other grasses where the coleoptile and scutellum become separated by what appears to be an internode this must be considered as a prolongation, by intercalary growth, of the scutellar node. This region has sometimes been called the mesocotyl, and differs structurally from all other internodes of the stem. Its method of elongation is also different from that of all other internodes in that the region of intercalary growth is at the upper end just below the attachment of the coleoptile, whereas in other internodes the growth region is basal, a little above the attachment of the leaf. There is also a tendency to form roots throughout the length of the first internode while in the other internodes they are limited to a whorl around the internode just above its base.

In some of the *Cyperaceae,* or sedges, which seem to be closely related to the grass family, the coleoptile is not present until after the leaves of the plumule have started development, and forms from the cotyledon during germination.

Those who are interested in the homologies of the grass embryo, a matter that is not likely to be definitely settled, should consult the following references: *(6, 7, 10, 11, 16, 56, 77, 93, 101).*

The growing point of the stem next begins to form leaf initials, the first of which is on the side of the stem opposite from the scutellum. The number formed varies from four to six, but five seems to be the usual number before the embryo becomes dormant in the mature seed (Fig. 64). Terminal growth in the leaf is soon replaced by basal growth. These leaves all remain rolled up inside of the coleoptile until emergence during germination.

At about the same time that the stem tip begins to differentiate, the tissues in the lower part of the embryo, just above the suspensor, begin differentiating to form the initial of the primary root. A number of lateral seminal root initials form just above the scutellar node, the number differing in different varieties of corn. Flint and flour corn often have but one root initial present in the embryo, while in dent varieties there are usually from three to five or more in the mature kernel.

The Mature Kernel

During the 50-day interval from double fertilization of the ovule to maturity of the grain (Figs. 65 and 66), the ovary with contents has increased in size from a nearly spherical body .03 inch in diameter to a slightly tapering kernel of screen size 13/64 x 25/64 x 32/64 inch (large, flat, cornbelt seed

Fig. 65.—Photograph. Ears of White Prize Dent corn harvested at weekly intervals before and after fertilization up to maturity. The age in weeks after seedling emergence is indicated below the ears. Meiosis occurred during the seventh week; fertilization took place during the eighth week, and the ears were mature (34 per cent kernel-moisture) by the end of the fifteenth week. X 0.2.

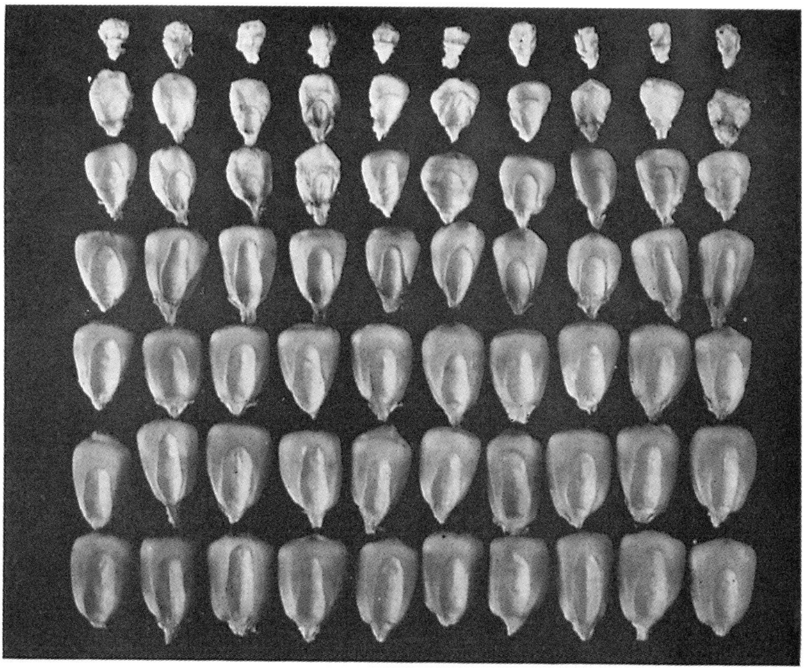

Fig. 66.—Photograph. Cured kernels of White Prize Dent corn harvested at weekly intervals (top to bottom) after fertilization at age of 8 weeks, until maturity at age of 15 weeks. These kernels were taken from the central portion of last seven ears shown in Figure 65. X 1.0.

grade). Their respective volumes are approximately .000014 and .02 cubic inch. This represents a volume increase of about 1400-fold. The ovary wall continues as the outer covering of the kernel. After a succulent growth by cell division and enlargement it has become transformed into the pericarp by disintegration and collapse of the interior region, thickening and lignification of the periclinal cell walls of the outer region, and finally compression of the entire tissue.

The integuments or seed coats which were fairly prominent in their initial stages have disintegrated to insignificant, scattered, non-cellular remnants. On the other hand, the outer wall of the nucellar epidermis has thickened and become suberized, and forms a continuous semi-permeable nucellar membrane between the aleurone and pericarp *(13, 32, 36,* and *71).* This extends to the protective closing layer *(36)* which lies completely across the placental-hilar region of the pericarp opposite the specialized absorbing cells of the endosperm at the base of the kernel. The disk of dark brown hilar tissue (for all practical considerations a hilum) is exposed to view only when the kernel breaks from the pedicel at its lower surface.

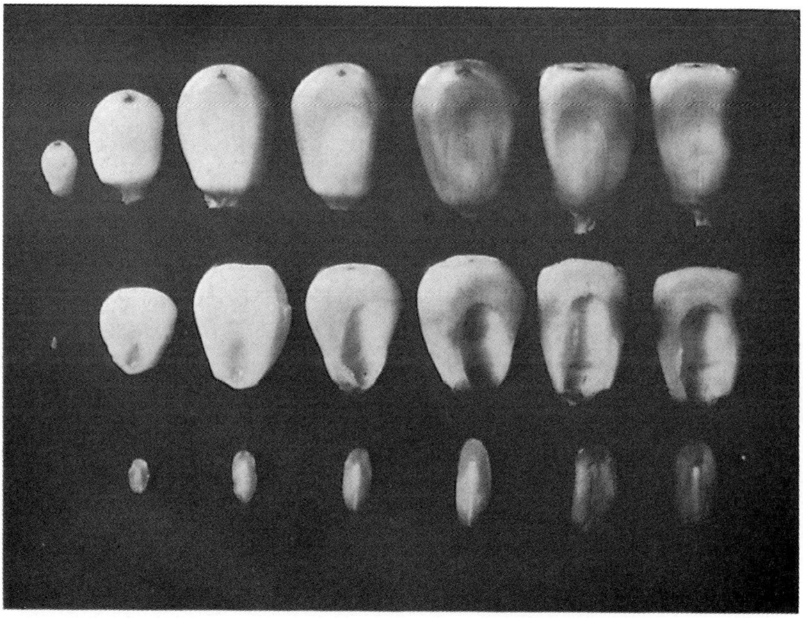

FIG. 67.—Relative size of kernel and its parts at successive weekly intervals, left to right, after fertilization to maturity. These kernels correspond with those shown in Figure 66. X 1.5.

Upper, kernel; middle, endosperm; lower, embryo.

The interior of the kernel is occupied by the embryo and endosperm. The endosperm constitutes about 85 per cent of the moisture-free weight of the mature kernel, the embryo 10 per cent, and the pericarp together with remnants of the nucellus and seed coats and the pedicel make up the remaining 5 per cent. Typical mature kernels of dent corn, together with the endosperms and embryos removed at weekly intervals after fertilization, are shown in Figure 67.

The embryo is the new sporophyte, whereas the pericarp and adhering nucellar membrane and pedicel are part of the mother plant of the previous sporophytic generation. The haploid gametophytic generation has almost or fully disappeared, a remnant of the antipodals sometimes persisting. The triploid endosperm is regarded as a continued development of the female gametophyte that is dependent upon the union of a sperm with the two polar nuclei in the process of double fertilization.

INHERITANCE IN CORN

FROM ANCIENT TIMES it had been noticed that the progeny of both plants and animals had a tendency to resemble their parents, but until recently ideas as to how this was brought about were vague and speculative. Real knowledge, based on experiments, has been mainly acquired since near the end of the last century when segregation of characters was noted independently by Correns, DeVries, and Tschermak. The discovery of Mendel's papers showed that he had determined the rules of segregation nearly 40 years earlier, but his results had remained practically unknown.

Genes

Based on the study of segregation in crosses and on other studies, the view is now general among geneticists that heredity is transmitted by genes, arranged in linear series in the chromosomes. Since in sporophytic tissue there are two sets of chromosomes, the genes generally occur in pairs, one derived from each parent. Both of the homologous genes making such a pair may be alike or they may be different. Such differences are supposed to have come about through a change or mutation of genes which were originally alike. When the genes of a pair are unlike, usually one is recessive and the other dominant. In case of complete dominance the recessive seems to have little or no effect, and the character determined by these genes develops the same as when both dominants are present. The recessive character appears only when both members of the pair are recessive. This pairing of recessives will happen at random in an average of 25 per cent of the fertilized eggs, the other 75 per cent being either homozygous dominant or heterozygous, with only one dominant. This will result in a 3:1 segregation in the next or second generation. Genes are not always inherited independently because of a tendency for linkage with other genes located in the same chromosome. Thus the genes occurring in a specific chromosome are said to belong to the same linkage group.

Some of the characters of the corn kernel are limited to the endosperm, such as yellow endosperm color, sweet or starchy endosperm, etc. As the endosperm receives two sets of chromosomes from the female parent and one set from the male parent, so in any set of genes there will be three instead of two. The two coming from the female parent will always be alike, and in some cases as when flint and flour corns are crossed the two genes for endosperm type will always dominate the single one. The crossed kernels will thus resemble the female parent in endosperm texture whichever way the cross is made. Which, if either, would be dominant if present in equal numbers is not known, but theoretically could be tested by using pollen from a plant with an extra chromosome carrying the gene for either the flint or flour character. In the case of incomplete dominance, as when white corn is crossed with yellow, the endosperm will show different shades of yellow, depending on the number of dominant genes present.

Some characters depend on the combined action of several different genes *(23 and 24)* as for instance purple aleurone color, which develops only in the presence of five dominant genes and one recessive. The genes which must be present as dominant are designated A_1, A_2, C, R, Pr, and the recessive is i. There may thus be several combinations producing white, some of which when crossed will produce red or purple corn. Inheritance is not determined by the characters shown by the parent phenotypes but by their gene composition or genotypes.

The study of chromosome morphology has not been undertaken as a part of these investigations. Among the investigators doing intensive research concerning the anatomy of the corn chromosomes, their behavior in meiosis, and the principles of cytogenetics in this species may be included Anderson *(3)*, Brink and Cooper *(12)*, Burnham *(14)*, Cooper and Brink *(19)*, Creighton and McClintock *(20)*, Longley *(55)*, McClintock *(57)*, Mangelsdorf and Reeves *(63)*, Randolph *(70)*, Rhoades *(73)*, Rhoades and McClintock *(74)*, Stadler *(80)*, and others. The mapping of genes in the various linkage groups of corn made striking headway under the leadership of R. A. Emerson and his associates. As many as 350 genes whose inheritance is known were listed in 1935 *(25)*, and many more have been reported since.

Xenia

Focke, who originated the term *xenia,* evidently intended to include all effects of foreign pollen which appeared in fruit, seed, or vegetative structure of the female parent plant. Any change in the tissue of the mother plant, such as change in weight or size of pericarp, is now called metaxenia. Change in embryo tissue, as in the color or composition of the cotyledons of peas, is not generally regarded as xenia, so that use of the term is now usually limited to any changes in the endosperm resulting as an immediate effect of foreign pollen *(92)*. The endosperm is sometimes called a xeniophyte.

A general term is needed that includes all direct, modifying effects of pollen on non-maternal tissues of the seed or caryopsis, whether endosperm or embryo. It would seem logical and satisfactory to have *xenia* serve in this broadened usage. This would simplify discussions which refer to the *xenia responses* of seeds, endosperms, and embryos, respectively.

Xenia is common in corn *(98* and *40)*, especially by way of aleurone- and endosperm-color responses and changes in the composition, texture and weight of the endosperm (Fig. 68). For example, the kernels of white sweet corn become yellow, starchy, and heavier when outcrossed by plants with yellow starchy kernels.

Xenia is very convenient in the demonstration of segregation, as the kernels can be counted, whereas most sporophytic responses are not apparent until the seed is planted and the character becomes visible in the progeny plants.

FIG. 68.—Xenia illustrated by ears that were fertilized with mixtures of pollen from the same and another variety or type.
1, yellow dent ear, hybrid kernels fertilized by blue flour.
2, yellow sweet ear, hybrid kernels fertilized by yellow dent.
3, yellow sweet ear, hybrid kernels fertilized by black sweet.
4, yellow pop ear, hybrid kernels fertilized by red pop.
5, white flint ear, hybrid kernels fertilized by yellow sweet.
6, white flint ear, hybrid kernels fertilized by yellow dent.

Some embryo responses may also appear in the crossed seed, such as purple coleoptile and colored scutellum, but have not been used as much as the endosperm characters to demonstrate segregation. Endosperm weight or yield responds to hybrid vigor in a manner rather similar to that of the embryo or sporophyte, though not necessarily in the same degree.

Maternal Inheritance

In some characters the inheritance seems to be determined by the female parent only, the pollen apparently having no effect. Some of the deficiencies of chlorophyll in corn due to abnormal chloroplasts belong to this class.

Two explanations have been suggested for maternal inheritance. Either the character is transmitted by the egg cytoplasm (the sperm contributing little if any cytoplasm) or, as in the case of defective plastids, these are transmitted from cell to cell, including the egg cell but not the very small sperm cell. By this means they are inherited independently of nuclei, chromosomes, and genes.

In some of the lower plants where the plastids function in both the gametophytic and sporophytic generations, such as mosses, liverworts, ferns, and fern allies, the plastids have been detected at all stages of plant growth. In *Selaginella* they have been found in the young spores forming in the spore mother cells. Such continuity has not been established in corn *(69)*.

Chromosomal Aberrations

In normal chromosome behavior in the gametophytes (embryo sac and pollen) each nucleus has a single set of chromosomes. This is the simple or haploid condition. When the sperm and egg nuclei unite upon fertilization, the resultant embryo has two sets and is diploid. When any interchanges of segments take place normally they are between the two homologous chromosomes.

Many variations from these conditions are found but in the breeding of corn they do not seem to be important as they do not generally give rise to new distinct forms. They have, however, proved useful in the study of the chromosomes and their behavior, such as crossing over, location of genes, etc. In *Oenothera* (evening primrose) and *Datura* (jimson weed) many of the striking forms are due to changes in chromosomes, such as increase in numbers, either by adding full sets or individual chromosomes. It was for such changes in the evening primrose that the term mutation was first applied by DeVries. One of these mutants, called gigas, was found to have double the normal number of chromosomes. In some of the cereals several groups based on chromosome number occur in which the larger numbers are multiples of the smaller as 7–14, 14–28, 21–42.

Polyploids have been found in corn, but the plants usually are not markedly different from normal plants, generally not enough different to be detected except by counting the chromosomes. When corn having 30 chromosomes is crossed with normal corn having 20, the extras are distributed irregularly during meiosis. As a result it has been possible to obtain plants with 21 (or 22) chromosomes in which any one of the 10 chromosomes has appeared as the extra.

Since the 10 chromosomes can be distinguished microscopically, it has been possible to determine which genes occur on each, as the group of genes occurring in the set with the extra chromosome will show a different segregation, giving trisomic ratios due to the presence of the extra gene. If there are two dominants and one recessive present instead of one of each there will be fewer homozygous recessives in the progeny with ratios of 35:1 instead of 3:1. Another type of aberration is the occurrence of translocations between chromosomes which are not homologous, i.e., not members of the same pair. Corn plants in which such a translocation has taken place may be perfectly fertile among themselves but show semisterility when crossed with normal corn. In such cases the chromosomes in pairing form unusual shapes as each segment pairs in synapsis with the corresponding segment of a normal chromosome.

In some cases of translocation, segments may be lost or an extra attached segment may be present. This leads to peculiar behavior in pairing with normal chromosomes. Such aberrations also have helped in locating the genes on chromosomes.

Inbreeding and Heterosis

When close bred, corn shows a reduction in vigor of growth and productivity which is most marked after self-fertilization. Various attempts have

FIG. 69.—Inbred lines derived from a standard local variety of dent corn shown at extreme right, after 12 generations of self-fertilization. Grown at Nebraska Agricultural Experiment Station in 1922.

FIG. 70.—First generation, single-cross hybrids between inbred lines shown in Figure 69. The yield of the best hybrids materially surpasses that of the original variety. Seed for the commonly grown commercial hybrid is produced by crossing together two such unrelated single crosses that have favorable specific combining ability. Grown at Nebraska Agricultural Experiment Station in 1922.

FIG. 71.—Second generation single-cross hybrids showing loss of productivity as compared with the F_1 hybrids in Figure 70. Grown at Nebraska Agricultural Experiment Station in 1922.

been made to explain this loss of vigor. The most generally accepted theory is that vigor is due to a large number of growth factors or genes, which in ordinary open-pollinated corn are present largely in a homozygous dominant or in a heterozygous condition, with relatively few present as homozygous recessives.

When selfed once, the homozygous pairs of genes will remain homozygous in the progeny, and about one-fourth of the heterozygous pairs become homozygous dominant, one-fourth homozygous recessive, and about half remain heterozygous. Assuming that vigor depends on dominants, such inbreeding greatly reduces vigor by virtue of increasing the number of homozygous recessives present. A corresponding reduction accompanies each successive selfing, and after about six times the corn becomes so homozygous that there is little further noticeable reduction in vigor (Fig. 69). The pairing among dominant and recessive genes is a matter of the law of chance, and each individual of the progeny is not reduced a half in heterozygosity by each selfing but the average of the population has its heterozygosity reduced a half. The individuals would fit in a probability curve with one limit as heterozygous as the original corn and the other limit at complete homozygosity; but the mode of greatest frequency would be at half the heterozygosity of the original plant that was inbred. A population of millions might be required for either limit to be reached and in a small random sample most of the individuals would rank close to the mode. Because of linkage between genes, it is not possible, theoretically, to obtain segregates that have only favorable or dominant growth factors for all plant characters.

Superior physiological functioning and anatomical and morphological organization, more resistance to disease organisms and insect pests, greater uniformity of favorable performance, and better adaptation to specific environmental conditions, are expressions of favorable gene combinations. In general, these dominant genes appear to contribute as much to the performance of the individual when present in heterozygous combination as when homozygous, but no more. Heterozygosity is not biologically an essential adjunct of vigor, but is an accompanying condition in the gene manipulation needed to establish maximum dominance.

As a result of repeated inbreeding some of the most deleterious recessives are removed by the plant becoming so reduced in vigor that it is eliminated by natural selection, and more may be eliminated by intelligent selection of only the most desirable plants for further inbreeding. Upon crossing the inbred lines, vigor is restored (Fig. 70), and the yields of some combinations surpass that of the original variety. However, only a small percentage of the lines produced prove useful in making outstanding hybrids. The breeding value of the lines must be established through repeated performance tests of hybrids in which they are involved. The prepotency or combining ability of an inbred line with respect to yield and other characters will depend upon both the number and the importance of the specific dominant genes which it contributes to its hybrid progeny. The most productive and desirable hybrids will be those with a maximum number of dominant or heterozygous factor pairs which have been selected for their favorable character expression. Of the enormous number of lines produced since Shull suggested the method of breeding corn by crossing inbred lines, relatively few are used in the production of commercial hybrid seed. The advantages of hybrid corn are largely restricted to the first generation progeny. The yield of a second generation single cross (Fig. 71) normally is about midway between that of the F_1 crop and the mean of its inbred parents. Corresponding yield losses in F_2 double–cross hybrids are only about half as great when the generation is advanced by sib–pollination.

Anatomically, about 10 per cent of the increase in plant size of hybrids over that of their inbred parents is due to increase in cell size and 90 per cent is due to greater cell numbers *(39)*. The hybrids are characterized by more rapid cell division during growth.

Corn Breeding

Breeding procedure.—In a modern corn breeding program *(35, 42,* and *79)* the usual sequence of procedures is as follows:

1. Development of inbred lines from open-pollinated varieties within and among which there is great variability (Fig. 72), by selection within self-fertilized lines. Improvement of lines may be achieved by backcrossing on hybrids and by inbreeding outstanding hybrids, accompanied by selection. Recurrent selection, designed to increase the frequency of favorable genes, also offers promise as a means for obtaining superior new lines.

2. Testing the prepotency or general combining ability of the more promising lines by the topcross or line-variety cross method. This consists of a thorough performance test of crosses between the respective lines and an ordinary open-pollinated variety. A modified procedure is to substitute as testers one or more hybrids with specific characters of known inheritance, for the open-pollinated variety.

3. Crossing in all possible combinations a limited number, up to about 15, of the lines with highest general combining ability. It is advantageous to include two superior established lines in such a group.

4. Testing for performance as to various characters all possible single-cross combinations among the selected lines in order to evaluate the lines for specific combining ability. (The formula to determine the number of single crosses possible from a given number (N) of lines $=\frac{(N-1)N}{2}$. Thus for 10 lines, 45 distinct single crosses are possible.)

5. Prediction of the performance of all possible double crosses based on the performance of the single crosses with respect to various agronomic characters. (The theoretical performance of any double cross equals the mean

FIG. 72.—Variations in botanical type, regional type, and varietal type of corn. All were grown comparably on the Nebraska Agricultural Experiment Station farm. Such natural variations provide a wealth of germ plasm for use in modern corn breeding.

performance of the four single crosses resulting from a combination of each of the lines of one single-cross parent with each of the lines in the other parent.)

6. Making seed of those double crosses which have the best predicted performance, and testing these preliminary to commercial production of the best. Such prediction eliminates the needless making and testing of many inferior hybrids. (The formula to determine the number of double crosses possible from a given number (N) of lines $= \frac{N\ (N-1)\ (N-2)\ (N-3)}{8}$. Thus for 10 lines, 630 distinct double crosses are possible.)

Hybrid seed production.—The double cross is the most common form of commercial hybrid seed corn. To a more limited extent seed of single crosses, three-way crosses, and topcrosses is produced for commercial planting.

For all practical considerations, reciprocal crosses are regarded as equal in their inheritance, and choice of which parent to use as female should be based largely upon vegetative and kernel characteristics which affect the economics of seed production and the appearance, size, or other quality characteristics of the seed.

In commercial seed production, hybridization is accomplished by planting the two parental stocks in alternate plots across the field and detasseling the seed parent which thereafter becomes cross pollinated by the tassel parent. Seed ears are picked only from the detasseled plants. Reasonable purity is provided by roguing off-type plants before flowering, and by reasonable isolation of the field from other corn. Should the use of male-sterile seed stocks become feasible in hybrid seed production, they would be included as the ear-parents in crossing fields, thereby eliminating the need for detasseling.

The possibility of establishing commercially feasible, superior, synthetic open-pollinated varieties by the combination of a large number of "high-combining" lines is still being explored.

The U. S. Department of Agriculture estimates (July 12, 1949) that 67,347,000 acres or 78 per cent of this country's corn acreage is planted to hybrid seed in 1949. In eight states, including Nebraska, the percentage is 94 per cent or more. This United States hybrid corn acreage compares with 0.1 per cent in 1933.

LITERATURE CITED

1. ANDERSON, EDGAR. Homologies of the ear and tassel of *Zea mays*. Ann. Mo. Bot. Gard. 31:325-342. 1944.
2. ———, AND BROWN, W. L. A morphological analysis of row number in maize. Ann. Mo. Bot. Gard. 35:323-336. 1948.
3. ANDERSON, ERNEST G. Chromosomal interchanges in maize. Genetics 20:70-83. 1935.
4. ARTSCHWAGER, ERNEST. Anatomy of the vegetative organs of sugar cane. Jour. Agr. Res. 30:197-221. 1935.
5. ———, BRANDES, E. W., AND STARRETT, RUTH C. Development of flower and seed of some varieties of sugar cane. Jour. Agr. Res. 39:1-30. 1929.
6. AVERY, GEORGE S., JR. Coleoptile of *Zea mays* and other grasses. Bot Gaz. 86:107-110. 1928.
7. ———. Comparative anatomy and morphology of embryos and seedlings of maize, oats, and wheat. Bot. Gaz. 89:1-39. 1930.
8. BONNETT, O. T. Development of the staminate and pistillate inflorescences of sweet corn. Jour. Agr. Res. 60:25-37. 1940.
9. ———. Ear and tassel development in maize. Ann. Mo. Bot. Gard. 35:269-287. 1948.
10. BOYD, LUCY. Evolution in the monocotyledonous seedling, a new interpretation of the grass embryo. Trans. Bot. Soc. Edinburgh 30:286-302. 1931.
11. ———, AND AVERY, GEORGE S., JR. Grass seedling anatomy, the first internode of *Avena* and *Triticum*. Bot. Gaz. 97:765-779. 1936.
12. BRINK, R. A., and COOPER, D. C. A proof that crossing over involves an exchange of segments between homologous chromosomes. Genetics 20:22-35. 1935.
13. BROWN, A. J. On the existence of a semipermeable membrane enclosing the seeds of some of the *Gramineae*. Ann. Bot. 21:79-87. 1907.
14. BURNHAM, C. R. Cytogenetic studies of an interchange between chromosomes 8 and 9 in maize. Genetics 19:430-447. 1934.
15. CAMPBELL, D. H. On the development of the stomata of Tradescantia and Indian corn. Amer. Nat. 15:761-766. 1881.
16. CELAKOVSKY, L. F. Über die Homologien des Grasembryos. Bot. Ztg. 55:141-174. 1897.
17. COLLINS, G. N. Notes on the agricultural history of maize. Ann. Rpt. Amer. Hist. Assoc. 1:409-429. 1919.

18. COOPER, D. C. Macrosporogenesis and embryo-sac development in *Euchlaena mexicana* and *Zea mays*. Jour. Agr. Res. 55:539-551. 1937.
19. ———, AND BRINK, R. A. Chromosome homology in races of maize from differing geographical regions. Amer. Nat. 71:582-587. 1937.
20. CREIGHTON, HARRIET B., AND MCCLINTOCK, BARBARA. A correlation of cytological and genetical crossing-over in *Zea mays*. Proc. Nat. Acad. Sci. 17:492-497. 1931.
21. CROZIER, A. A. Silk seeking pollen. Bot. Gaz. 13:242. 1888.
22. CUTLER, H. C., and CUTLER, M. C. Studies on the structure of the maize plant. Ann. Mo. Bot. Gard. 35:301-316. 1948.
23. EAST, E. M. Inheritance of color in the aleurone cells of maize. Amer. Nat. 46:363-365. 1912.
24. EMERSON, R. A. Latent colors in corn. Amer. Breeders Assoc. Rpt. 6:233-237. 1911.
25. ———, BEADLE, G. W., AND FRASER, A. C. A summary of linkage studies in maize. Cornell Agr. Exp. Sta. Memoir 180. 81 pp. 1935.
26. EVANS, ARTHUR T. Vascularization of the node in *Zea mays*. Bot. Gaz. 85:97-103. 1928.
27. FISK, EMMA L. The chromosomes of *Zea mays*. Amer. Jour. Bot. 14:53-75. 1925.
28. GAGER, C. STUART. An account of the glands in the endosperm of *Zea mays*. Bul. Torrey Bot. Club 34:125-134. 1907.
29. GOLINSKI, S. J. Ein Beitrag zur Entwicklungsgeschichte des Androeceum und des Gynaeceum der Graser. Bot. Centbr. 55:1-135. 1893.
30. GORDON, MARY. The development of the endosperm in cereals. Proc. Roy. Soc. Victoria 34:105-116. 1922.
31. GUIGNARD, L. La double fecundation dans le mais. Jour. Botanique 15:37-50. 1901.
32. HARRINGTON, G. T., AND CROCKER, WILLIAM. Structure, physical characteristics, and composition of the pericarp and integument of Johnson grass seed in relation to its physiology. Jour. Agr. Res. 23:193-222. 1923.
33. HARZ, C. O. Zea. Landwirtschaftliche Samenkunde. pp. 1237-1243. 1885.
34. HAYS, W. M. Indian corn, habits of root growth, etc. Minn. Agr. Exp. Sta. Bul. 5:5-8. 1889.
35. JENKINS, M. T. Corn improvement. U. S. Yearbook of Agriculture, pp. 455-522. 1936.
36. JOHANN, HELEN. The histology of the caryopsis of yellow dent corn, with reference to resistance and susceptibility to kernel rots. Jour. Agr. Res. 51:855-883. 1935.
37. KEMTON, J. H. Floral abnormalities in maize. U. S. D. A. Bul. 278. 16 pp. 1913.
38. KIESSELBACH, T. A. Transpiration as a factor in crop production. Nebr. Agr. Exp. Sta. Res. Bul. 6. 214 pp. 1916.
39. ———. Corn investigations. Nebr. Agr. Exp. Res. Bul. 20. 152 pp. 1922.
40. ———. The immediate effect of gametic relationship and of parental type upon the kernel weight of corn. Nebr. Agr. Exp. Sta. Res. Bul. 33. 69 pp. 1926.
41. ———. Fasciated kernels, reversed kernels, and related abnormalities in maize. Amer. Jour. Bot. 13:33-39. 1926.
42. ———. The possibilities of modern corn breeding. Proceedings of the World Grain Exhibition and Conference, Canada. Vol. 2:92-112. 1933.
43. ———. Effects of age, size, and source of seed on the corn crop. Nebr. Agr. Exp. Sta. Bul. 305. 16 pp. 1937.
44. ———, AND KEIM, F. D. The regional adaptation of corn in Nebraska. Nebr. Agr. Exp. Sta. Res. Bul. 19. 64 pp. 1921.
45. ———, AND PETERSEN, N. F. The chromosome number of corn. Genetics 10:80-85. 1925.
46. ———, AND RATCLIFF, J. A. Freezing injury to seed corn. Nebr. Agr. Exp. Sta. Res. Bul. 16. 96 pp. 1918.
47. KUWADA, Y. Meiosis in the pollen mother cells of *Zea mays*. Bot. Mag. Tokyo 25:163-181. 1911.
48. ———. Die chromosomenzahl von *Zea mays* L. Jour. Col. Sci. Imp. Univ. Tokyo 39:1-148. 1919.
49. ———. On the number of chromosomes in maize. Bot. Mag. Tokyo 39:227-234. 1925.
50. LAMPE, LOIS. A microchemical and morphological study of the developing endosperm of maize. Bot. Gaz. 91:337-376. 1931.

51. LENZ, L. W. Comparative histology of the female inflorescence of *Zea mays* L. Ann. Mo. Bot. Gard. 35:353-376. 1948.
52. LAUBENGAYER, R. A. The vascular anatomy of the four-rowed ear of corn. Ann. Mo. Bot. Gard. 35:337-340. 1948.
53. LINDSTROM, E. W. Heritable characters of maize. A new dominant hereditary character—Teopod. Jour. Hered. 16:135-140. 1925.
54. LONGLEY, ALBERT E. Chromosomes in maize and maize relatives. Jour. Agr. Res. 28:673-682. 1924.
55. ———. Chromosome morphology in maize and its relatives. Bot. Rev. 7:263-289. 1941.
56. MCCALL, M. A. Developmental anatomy and homology in wheat. Jour. Agr. Res. 48:283-321. 1934.
57. MCCLINTOCK, BARBARA. Chromosome morphology in *Zea mays*. Science 69:629. 1929.
58. ———. A cytological demonstration of the location of an interchange between two non-homologous chromosomes of *Zea mays*. Proc. Nat. Acad. Sci. 16:791-796. 1930.
59. ———. Cytological observations of deficiencies involving known genes, translocations, and an inversion in *Zea mays*. Mo. Agr. Exp. Sta. Bul. 163. 78 pp. 1931.
60. ———. The association of non-homologous parts of chromosomes in the mid-prophase of meiosis in *Zea mays*. Z. Zellf. Mik. Anat. 19:191-237. 1933.
61. MAGEE, J. A. Histological structure of the stem of *Zea mays* in relation to stiffness of stalk. Iowa State Coll. Jour. Sci. 22:257-268. 1948.
62. MANGELSDORF, P. C. The role of pod corn in the origin and evolution of maize. Ann. Mo. Bot. Gard. 35:323-336. 1948.
63. ———, AND REEVES, R. G. The origin of Indian corn and its relatives. Texas Agr. Exp. Sta. Bul. 574. 315 pp. 1939.
64. MARTIN, JOHN N., AND HERSHEY, L. Ontogeny of the maize plant. Iowa State Coll. Jour. Sci. 9:489-502. 1934.
65. MILLER, E. C. Development of the pistillate spikelet, and fertilization in *Zea mays*. Jour. Agr. Res. 18:255-266. 1919.
66. MONTGOMERY, E. G. What is an ear of corn? Pop. Sci. Monthly 68:55-62. 1906.
67. PIPER, C. V. The terminology of the parts of the grass spikelet. Science 23:789-790. 1906.
68. RANDOLPH, FANNIE RANE. A cytological study of two types of variegated pericarp in maize. Cornell Agr. Exp. Sta. Memoir 102. 14 pp. 1926.
69. RANDOLPH, L. F. Cytology of chlorophyll types in maize. Bot. Gaz. 73:337-375. 1922.
70. ———. Chromosome numbers in *Zea mays*. Cornell Agr. Exp. Sta. Memoir 117. 44 pp. 1928.
71. ———. Developmental morphology of the caryopsis in maize. Jour. Agr. Res. 53: 881-916. 1936.
72. REEVES, R. G. Chromosome studies in *Zea mays*. Proc. Iowa Acad. Sci. 32:171-175. 1925.
73. RHOADES, M. M. A cytologic study of a chromosome fragment in maize. Genetics 21:491-502. 1936.
74. ———, AND MCCLINTOCK, BARBARA. The cytogenetics of maize. Bot. Review 1: 292-325.
75. RHOADES, VIRGINIA H. A study of fertilization in *Zea mays*. Thesis, Cornell Univ. 1934.
76. ROWLEE, W. W., AND DOHERTY, M. W. The histology of the embryo of Indian corn. Bul. Torrey Bot. Club 25:311-315. 1898.
77. SARGENT, E., AND ROBERTSON, AGNES. The anatomy of the scutellum in *Zea mays*. Ann. Bot. 19:115-122.
78. SCHUSTER, JULIUS. Über die Morphologie der Grasblute. Flora 100:213-266. 1909.
79. SPRAGUE, G. F. The experimental basis for hybrid maize. Biol. Rev. 21:101-120. 1946.
80. STADLER, L. J. On the genetic nature of induced mutations in plants. II. A haplo-viable deficiency in maize. Mo. Agr. Exp. Sta. Res. Bul. 204. 29 pp. 1933.
81. STEPHENS, S. G. A comparative study of a dwarf mutant in maize, and its bearing on the interpretation of tassel and ear structure. Ann. Mo. Bot. Gard. 35:289-299. 1948.
82. STEWART, ALBAN. The pistillate spikelet of *Zea mays*. Science 42:694. 1915.
83. STRASBURGER, EDWARD. Ein Beitrag zur entwicklungs Geschichte der Spaltoffnungen. Jahr. Wiss. Botanik 5:297-336. 1866.

84. STURTEVANT, E. L. The superabundance of pollen in Indian corn. Amer. Nat. 15:1000. 1881.
85. ———. Notes on maize. Bul. Torrey Bot. Club 21:319-343. 1894.
86. TRUE, RODNET H. On the development of the caryopsis. Bot. Gaz. 18:212-226. 1893.
87. VAVILOV, N. I. Studies on the origin of cultivated plants. Bul. Appl. Bot. Gen. and Pl. Breeding 16:139-248. 1926.
88. WALKER, ELDA R. On the structure of the pistils of some grasses. Univ. Nebr. Studies 6:203-218. 1906.
89. WEATHERWAX, PAUL. Morphology of the flowers of *Zea mays*. Bul. Torrey Bot. Club 43:127-144. 1916.
90. ———. The development of the spikelets of *Zea mays*. Bul. Torrey Bot. Club 44:483-496. 1917.
91. ———. The evolution of maize. Bul. Torrey Bot. Club 45:309-342. 1918.
92. ———. Gametogenesis and fecundation in *Zea mays* as the basis of xenia and heredity in the endosperm. Bul. Torrey Bot. Club 46:73-90. 1919.
93. ———. The homologies of the coleoptile and position of the scutellum in maize. Bot. Gaz. 69:73-90. 1920.
94. ———. The story of the maize plant. Univ. of Chicago Press. 247 pp. 1923.
95. ———. Persistence of the antipodal tissue in the development of the seed of maize. Bul. Torrey Bot. Club 53:381-384. 1926.
96. ———. The endosperm of Zea and Coix. American Jour. Bot. 17:371-380. 1930.
97. ———. The phylogeny of *Zea mays*. Amer. Midland Nat. 16:1-71. 1935.
98. WEBBER, H. J. Xenia, or the immediate effect of pollen in maize. U. S. D. A. Div. Veg. Phys. and Path. Bul. 22. 44 pp. 1900.
99. WEIHING, R. M. The comparative root development of regional types of corn. Jour. Amer. Soc. Agron. 27:526-537. 1935.
100. WIGGANS, ROY G. The number of temporary roots in the cereals. Jour. Amer. Soc. Agron. 8:31-37. 1916.
101. WORDSELL, W. E. The morphology of the monocotyledonous embryo and that of the grass in particular. Ann. Bot. 30:509-524. 1916.
102. YOUNG, PHOON T. Histogenesis and morphogenesis in the primary root of *Zea mays*. Thesis, Columbia Univ. 40 pp. 1933.

Index